ABBEVILLE PRESS

ENCYCLOPEDIA OF NATURAL SCIENCE

T*he World of Mammals* is one of the first five volumes in Abbeville Press's outstanding new collection of pocket encyclopedias of natural science. Written in scholarly yet easy-to-understand language, the series is packed with information that will fascinate student and nature buff alike. Beautiful full-color illustrations on every page supplement the lively text.

The World of Mammals is a comprehensive study which examines the evolution and development of mammals and the ways in which they have adapted to a wide variety of ecological threats. The volume combines the usefulness of a reference work with the readability of a browsing book, perfect for anyone attuned to the splendor of our natural world. It both explains and illustrates how mammals care for their young, how they survive, where they live—and more.

Together with *The World of Birds, Fish, Insects,* and *Amphibians and Reptiles,* these books form the cornerstone of an indispensable home library—one that is almost guaranteed not to sit idly on the shelf.

The World of
MAMMALS

BY AUGUSTO VIGNA TAGLIANTI
TRANSLATED FROM THE ITALIAN BY John Gilbert

ABBEVILLE PRESS • PUBLISHERS • NEW YORK

Designers/Artists
Illustrations drawn by Piero Cattaneo, Bergamo, Piero Cozzaglio, Brescia, and Remo Squillantini, Florence, with the exception of the following: Giorgio Arvati, Verona: 102-103; Luciano Corbella, Milan: 26, 248; Alesandro Fedini, Milan: 20-21, 48; Raffaello Segattini, Verona: 11, 12-13, 29, 41, 74, 90, 91, 203.

Photographs
Archivio Mondadori, Milan: 19, 189 above left, 234; Walter Bonatti: 57 bottom; Ardea Photographics, London, I.R. Beames: 131; British Museum, London: 15; F. Carrese, Milan: 63; © Walt Disney Productions, Burbank, California, U.S.A.: 89; Esso Standard Italiana, Rome: 191; Feverstein-Schuol Schuls, Switzerland: 129; Grant: 83a; C. M. Hladik, Paris: 75; Jacana, Paris, Bel et Vienne: 42, 215, 249; Bassot: 205, 231; Bos: 62; Brosselin: 92, 159, 161; Chaumeton: 51, 70, 71; Dannegger: 158; Devez: 134-135b, 136l, 152; Fievet: 83b, 166b; Frédéric: 66; Fritz: 73; Gerard: 39; Grossa: 178: Haüsle: 110, 163; Lelo: 99; Letellier: 144; Milwaukee: 50, 149; Peter: 236; Renaud: 204; Robert: 58-59, 98; Roux: 33; Schraml: 31, 40 right, 125; Solaro: 219b; Stoll: 46; Simm: 77; Sundance: 179; Vatin: 166a; Visage: 40l, 61, 65, 80, 143, 181; Wangi: 32; A. Jolly: 138; Aldo Margiocco, Genoa: 95b, 189b; Giuseppe Mazza, Milan: 45, 134a, 228-229; Lino Pellegrini, Milan: 112, 224-225; Pictol, Milan: 49, 68-69, 88a, 145, 199; Luisa Ricciarini, Milan, Carlo Bevilacqua: 136r; G. Capelli: 95a, 165b; Nino Cirani: 44, 242; P. Curto: 230; Angelo P. Rossi: 81, 226a; Gustavo Tomsich: 56, 84; Fulvio Roiter, Venice: 64, 86-87; A. Schilling, Paris: 139; Ariberto Segala: 30; Union Press, Milan: 93.

Printed and bound in Italy by Officine Grafiche of Arnoldo Mondadori Editore, Verona.

LC 80-69173

ISBN 0-89659-183-2
ISBN 0-89659-184-0 pbk.

Contents

Introduction

To read the story of the mammals is to discover one of Nature's greatest successes. Except for the polar ice caps and the deepest parts of the oceans, mammals have made their homes over the whole surface of the globe and are to be found in deserts and on mountains as well as in more hospitable lowland plains and forests. There are about 4,230 different kinds of mammals and although some of these have a restricted distribution and are therefore not numerous, some species have existed in immense numbers in the past. The herds of bison in North America were reckoned to number over 60 million animals before the coming of the white man and even today the wood mouse (*Apodemus sylvaticus*) is thought to be the most abundant mammal in Britain, outnumbering by a considerable margin the 54 million humans who share the environment with them.

The success of the mammals is due partly at least to the fact that they are active, warmblooded creatures, maintaining a steady body temperature and generally guarding against heat loss with an insulating coat of hair or fur. Infant mammals are nourished in a unique way, for their mothers produce milk, a food

specially adapted to their early needs. During the period of suck-ling, a strong bond may be forged between mothers and offspring and in many species this is maintained in later life. Some, particularly those which are fairly long lived, may be social and the young are educated in the ways of their kind so that their behaviour includes a wide range of learned as well as innate responses to the problems of survival. This adaptability is best seen in man, the most abundant large mammal, who occupies every corner of the earth and whose ingenious mind has invented machines to perform activities which other animals can undertake only by means of physical action.

Structural variability in mammals at least equals mental adaptability, and their body plan can be the basis for a wide range of specialisations to suit many ways of life. Other groups of animals can usually be identified by their overall shape, but mammals such as bats can fly, using wings formed by their modified front limbs. They have often been mistaken for birds, which use a similar method of flight, but they are, in every detail of their anatomy and physiology, true mammals.

▲ Long-eared bat (*Pleatus amitus*)

The same way may be said for the whales and dolphins, which are streamlined swimmers, often thought to be fishes, but again they are true mammals. A more familiar adaptation is the lengthening of all of the limbs to enable the creature to escape from danger by running fast, a way of life found in the deer, antelopes and horses.

Along with the variety in their ways of life, mammals differ in their feeding habits. They may be plant eaters, or herbivores, in some cases feeding on a wide range of green foods, in others, a few types of leaves only. The koala, for instance, will consume nothing but the young leaves of a small number of species of eucalyptus trees. Some mammals are flesh eaters, or carnivores, but again the variety of diet is very large, ranging from the flesh of freshly killed large animals to the last remains of a corpse taken

▼ Common dolphin
(*Delphinus delphis*)

by carrion feeders. Some, known as piscivores, feed on fishes; others, called insectivores, feed on insects and other kinds of invertebrates, while some bats feed only on blood. A very few species, including man and the pigs, feed on almost anything, and are referred to as omnivores.

With the possibility of living and feeding in so many ways, it might be thought that mammals have the world at their feet, to occupy all places as they please. This is not so, for many species are restricted to certain parts of the world, although there is no reason why they should not flourish elsewhere. In fact they often do so when transported by man and the spectacular success of the rabbit in Australia is an example of an animal taken from its native habitat to a climatically suitable region far from its place of

▼ Fallow deer (*Dama dama*)

origin. Many other animals owe their present distribution to man. In Britain, the rabbit, black and brown rats, the coypu, grey squirrel, mink and several species of deer were all brought from other parts of the world by human agencies. On a broader scale, however, the pattern of mammal distribution reflects the area of origin of various Orders and Families. Successful species often migrated to other parts of the world, sometimes into areas subsequently isolated by changes in land and sea level or by continental drift. An example may be seen in the pouched mammals (map p.108). These originated in North America in the Late Cretaceous Period, about 95 million years ago. At that time there was a land bridge between North America and Europe enabling a few marsupials to migrate eastwards. Some went south, into a great continent, part of which tore away and drifted off to become Australia. In the meantime, placental mammals, in which there is a prolonged prenatal connection between the mother and her young, developed in the Old World. These quickly ousted the less intelligent and less reproductively efficient marsupials in Europe and North America and a few of them got through to South America (although the land joining the two continents sank beneath the sea, forming a barrier which most could not cross), while none reached the island continent of Australia. North America then split from Europe, but much later a land link between North and South America was re-established and mammals moved across it in both directions. This was disastrous for the southern marsupials, which had evolved in isolation into many strange forms which became extinct in competition with the invading placentals. However, some of the original South American mammals survived and a very few of them moved north successfully. The most remarkable is the Virginian opossum, which is now found throughout much of North America as far as the Canadian border.

A nother geological occurrence which has determined the distribution of many mammals is the recent Ice Age, which for much of the last million years covered a good deal of the northern hemisphere with ice sheets over 1,000m thick. The ice was formed of water taken from the sea, so sea levels were lowered during ice advances and raised in interglacial phases. Land bridges, formed during periods of low sea level, allowed mammals into hitherto uncolonised areas, where they were sometimes cut off as the ice retreated and the sea advanced. In Britain, which was depleted of many mammals during the last phase of the Ice Age, the rapid rise in sea level as the ice melted prevented many mammals from entering the newly warmed land and Britain has a much poorer fauna than mainland Europe. Thus, where mammals live may be due to human, biological or geological factors.

In a purely structural sense, mammals show their relationships to many other creatures. At the most basic level, this can be seen in the possession, at some time in their lives, of gills and a stiffened, jellylike rod, called the notochord, which runs down the back with the main nerve cord above it and the gut below. In adult mammals the notochord is totally replaced by the vertebrae or backbones, but the possession of a notochord by many creatures, even some invertebrates, such as the 'tadpole' larvae of sea squirts tells of a relationship which is far from clear at first. An animal in which it is more obvious is Amphioxus, the Lancelet, which looks like a small fish, but which retains its notochord throughout life.

Fossils of the earliest chordates (animals with a notochord) are found in rocks of middle Cambrian age, dating back about 550 million years, but the first creatures with true bone do not occur until much later in the geological record, about 480 million years ago. In these animals the bone was external, forming a heavy armour, covering their bodies. Their internal support was cartilaginous, as it is in their descendants, the small number of jawless fishes of today. The origin of all of the groups of early vertebrates is obscure; the armoured fishes or Placoderms, which had internal as well as external bone, occur first

▼ (1) Larva of urochord; (2) of lancelet; (3) of a vertebrate

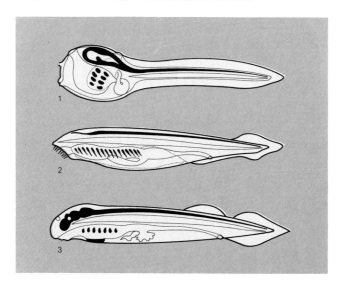

in the latest Silurian, about 430 million years ago and shortly
after, sharks and true fishes, which flourish to the present day,
developed. During the Devonian Period, about 380 million years
ago the first amphibians evolved from fishes which had the
ability to breathe dry air and to support themselves with heavy
fins when the water in which they were living in swamps and
desert oases dried up. Reptiles developed in the Carboniferous

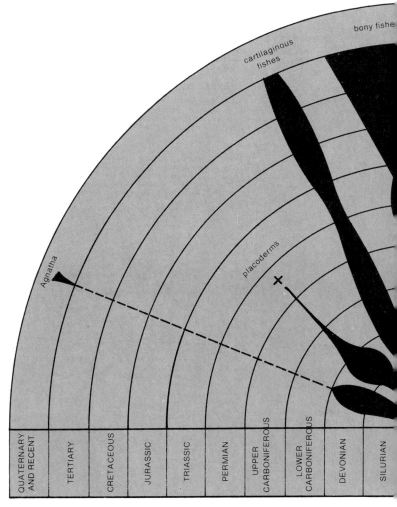

▲ Evolutionary chart of the vertebrates

Period, about 345 million years ago, and from separate branches of this great class, the mammals evolved about 200 million years ago in the Triassic Period and the birds in the Jurassic, about 170 million years ago.

Today the vertebrates have evolved so that the fishes can only occupy water. Amphibians can move on dry land, but are tied to water by their fish-like reproductive pattern. Reptiles,

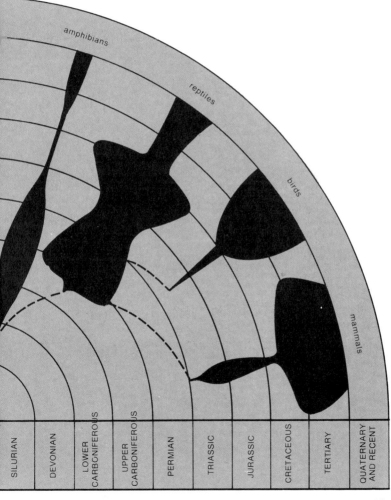

with their dry skin and shelled eggs are largely free of water, but their cold-blooded body system means that they can flourish only where the climate is fairly warm. Mammals, with their high activity and good body insulation, can occupy the colder parts of the earth; while birds, with the power of flight, can reach isolated areas inaccessible to land mammals.

There is a foreshadowing of the mammal condition in very early reptiles, such as the sail-back, *Dimetrodon*, which lived about 280 million years ago. This creature had a skull structure similar to that of the mammal ancestors and its way of life was probably much more active than that of its contemporaries. Two features in particular tell of this. One is the 'sail' supported by long, slender bones arising from the vertebrae. This enlarged its area without increasing its bulk much. Working like a two-way radiator, it allowed the animal to absorb heat when it was cold, or to blush off excess warmth, going a long way to stabilising

▼ Skeleton of *Dimetrodon*, synapsid pelycosaur

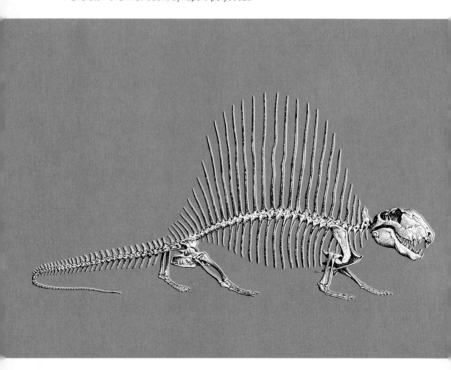

temperature and hence metabolic activity. The other feature denoting activity is the rather slender limb bones, suggesting the possibility of fairly speedy movement. Also the teeth show some specialisation over the usual reptile condition, for there are small nipping and holding teeth in the front of the mouth, tearing teeth at the corners of the jaws and behind these, a row of teeth for slicing food. It is a development of this type of tooth structure which can be seen in the mammal-like reptiles, such as *Cynognathus* where incisors, canines and molar type teeth, similar to those of modern mammals have developed.

The illustrations on the pages overleaf show the development of improved locomotion in the mammals as compared to their ancestors. The mammal-like reptile *Diademodon* had fairly slender limbs which raised the body clear of the ground at all times and which could be moved in a fore and aft fashion, rather than in the sprawling manner typical of reptiles. In modern

▼ Skull of *Cynognathus*, synapsid therapsid

15

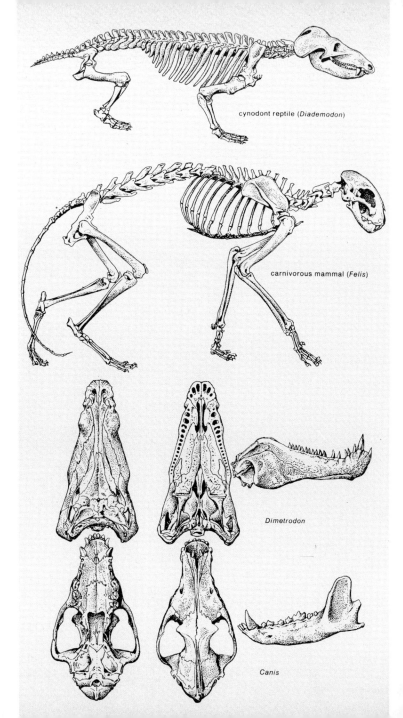

cynodont reptile (*Diademodon*)

carnivorous mammal (*Felis*)

Dimetrodon

Canis

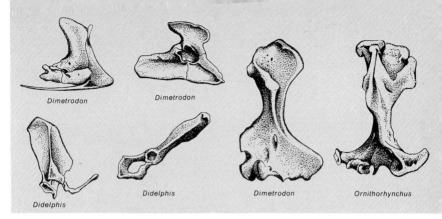

▲Evolution of the
shoulder girdle

▲Evolution of the
pelvic girdle

▲Evolution of the
humerus

mammals, such as cats, the limbs have become much longer and
slenderer and there is no hint of the reptile sprawl. The vertebrae
form a more complex girder to support the body in vertical as well
as lateral movement, so that galloping, a very efficient fast gait,
becomes possible. The shoulder and hip girdles of ancient reptiles
are heavy compared to those of modern mammals, such as the
opossum, which shows a typical slenderness of limb bones, with
complex articulations which allow precise movements. Fast
moving animals need more fuel, in the form of food, than is
required by sluggish creatures. One of the differences between
reptiles and mammals is that in the latter digestion begins much
more effectively in the mouth, with the food being broken up by a

▼ Evolution of the femur

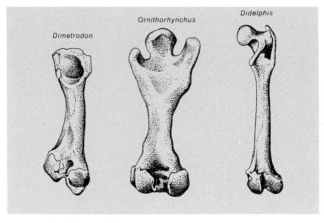

◄ Comparison between the skeleton, skull and
mandible of a synapsid reptile and a mammal

small number of teeth which are highly specialised in function, and then mixed with saliva, the first of the digestive juices. An active animal requires more oxygen than an inactive one and there are indications that regular, mammal type of breathing began with the mammal-like reptiles, in which a shelf of bone forming a secondary palate grew across the roof of the mouth to separate completely air breathed in by the nostrils from food in the process of mastication. Changes in the skull included a

▲Skeleton of a Miocene marsupial (*Prothylacynus*)

general reduction of bone to lighten the weight of the head and to allow areas where heavy muscles could bulge as the animal chewed. These enlarged chewing muscles required extra attachment areas on the lower jaw and an upward extension of the jaw bone developed. The presence of this demanded a new type of hinge by which the back of the jaw could articulate with the skull (see diagram p. 26). When dealing with ancient fossils, the disappearance of certain bones, characteristic of the reptile jaw hinge is the diagnostic feature by which mammals are identified, for none of the soft parts by which modern mammals are recognised are present and the first mammals show little of the huge brain extension which is the hallmark of their modern descendants.

The first true mammals developed at about the same time as the earliest of the dinosaurs, but they remained small and probably insignificant members of the total fauna for about 130 million

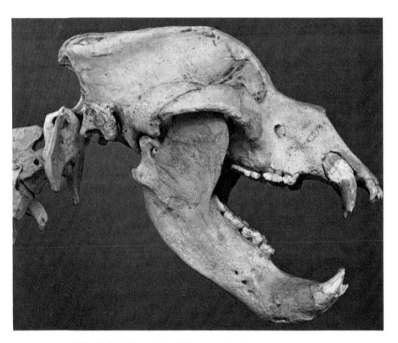

▲ Skull of giant cave bear (*Ursus spelaeus*)

years. Towards the end of this time the placental mammals evolved and were ready to take over the environment as the last dinosaurs became extinct. Three main groups of mammals now exist. These are the small and primitive 'living fossils' – the egg-laying Duckbilled platypus and spiny anteater. Second are the marsupials or pouched mammals, which produce their young at a very early stage of development and then nourish them in a pouch for a period which about equals the pregnancy of a placental

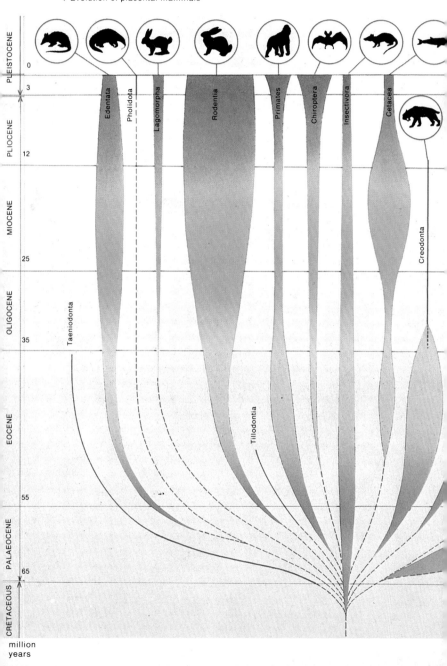

▼ Evolution of placental mammals

PLEISTOCENE

0

3

PLIOCENE

12

MIOCENE

25

OLIGOCENE

35

EOCENE

55

PALAEOCENE

65

CRETACEOUS

million
years

Edentata

Pholidota

Lagomorpha

Rodentia

Primates

Chiroptera

Insectivora

Cetacea

Creodonta

Taeniodonta

Tillodontia

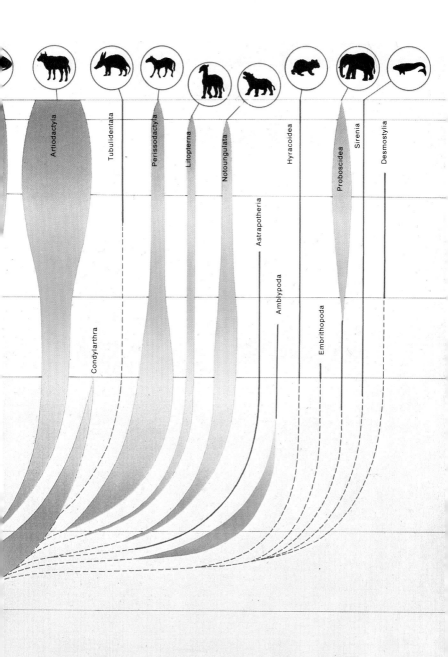

Artiodactyla

Tubulidentata

Perissodactyla

Litopterna

Notoungulata

Astrapotheria

Amblypoda

Hyracoidea

Embrithopoda

Proboscidea

Sirenia

Desmostylia

Condylarthra

mammal of a comparable size. The placentals, in which there is a special organ, the placenta through which the young are nourished before birth, are more specialised in many respects and are far more intelligent – compare the enlarged brain case of the cave bear with that of the fossil marsupial *Prothylacinus*.

From an insectivore-like ancestor, the placental mammals radiated rapidly to occupy many niches in the early Tertiary environment. Some like the Tillodonts and Embrithopoda, an Order known solely by one genus recorded only from the Fayum in Egypt, seem to have been evolutionary misfits, and were perhaps too specialised to survive competition and changes in habitat even at that early time. Others, such as the Condylarthrans, gave rise to a complex of other Orders which finally outstripped them in the evolutionary race and they became extinct at the end of the Eocene Period, about 35 million years ago. The zenith of the age of mammals was probably about 12 million years ago, when there were many more species than at present. The only groups which seem to be expanding today, as can be seen from the diagram pp. 20–21, are the primates and the bats, but this is probably illusory since the primates, as mainly tree dwelling creatures, stand a poor chance of becoming fossilized and our knowledge of monkey-like animals of the past is likely to be very incomplete. Bats have, on the whole, such fragile skeletons that they rarely become well fossilized, so again we have an inadequate record. However, at the end of the Pliocene Period there were many species of, for example, elephants and rhinoceroses, many of which were widespread, as well as of gigantic size. Many other Orders showed a greater diversity then than they do today. A large number of species of mammals and several Orders became extinct during or just after the last Ice Age and today there are only 16 placental Orders. Several of these are small, consisting of few species, with a restricted distribution and way of life. Examples are the flying lemurs (Dermoptera) and the aardvark (Tubulidentata). The most successful are the rodents, for nearly half of all kinds of living mammals belong to this Order. Although they are mostly of small size they have colonised a very wide range of habitats and among their ranks are burrowers, gliders, climbers and swimmers. Some of them are hugely successful in their way of life, and some species may contain very large numbers of individuals. Less varied in their ways of life, for all are flyers, are the bats, which form nearly a quarter of the total of mammal species. However, many of the smaller groups show great diversity and vitality and if man, a recently evolved species among them, can control his destructive activities, they will remain a dominant life form for many aeons to come.

The orders of mammals (extinct orders in italics) ▶

Infraclasses	Orders			
	Monotremata			
	Multituberculata	*Docodonta*	*Triconodonta*	
PANTOTHERIA	*Pantotheria*	*Symmetrodonta*		
METATHERIA	Marsupialia			
EUTHERIA	Insectivora	Dermoptera	Chiroptera	*Tillodontia*
	Taeniodonta	Edentata	Pholidota	Lagomorpha
	Rodentia	Cetacea	Carnivora	*Condylarthra*
	Lipoterna	Notoungulata	Astrapotheria	Tubulidentata
	Pantodonta	Dinocerata	Pyrotheria	Proboscidea
	Embrithopoda	Hyracoidea	Sirenia	Perissodactyla
	Artiodactyla	Primates		

Structure

Distinctive characteristics

As discussed in the introductory chapter, the principal features which distinguish mammals from other animal classes consist of a constant body temperature, a covering of hair and the presence of mammary glands. With the exception of the egg-laying monotremes, such as the Duckbilled platypus, mammals give birth to live young. The former also differ from other mammals in having a cloaca instead of separate anal and urinogenital openings. Except for the monotremes and the majority of marsupials, the mammalian embryo is nourished by means of a placenta.

Differences in skeletal structure between mammals and reptiles are of vital importance to the palaeontologist studying animal evolution on the basis of fossil remains. There are, however, also variations in the skeletal structure of mammals themselves. Most mammals, for example, have seven cervical vertebrae but some have only six, while others have as many as nine or ten. Most mammals have four limbs but in some, such as the cetaceans, there are only two forelimbs transformed into flippers. There are also variations in the number of toes. The shoulder girdle may or may not be equipped with a clavicle, and in the case of the monotremes it is of the reptilian type, with an interclavicle and coracoid bone. The teeth of most mammals are heterodontic so that the various types of teeth such as incisors, canines and molars are adapted to perform special functions. The exceptions

include dolphins whose teeth are all of equal shape, while other groups of whales and some ant-eaters have no teeth at all. A set of milk teeth is generally replaced by a permanent set.

The most revealing differences, however, relate to the structure of certain parts of the skull, and the most important distinguishing feature between reptiles and mammals is the presence or absence of the quadrate and articular bones. The illustration on p. 26 shows that in the ear region, there is a hinge between the mandible (lower jaw) and the skull. In mammals the mandible articulates directly with the squamous bone of the skull. By contrast, all reptiles and birds possess a variably sized articular bone which forms a hinge with the quadrate bone of the upper jaw. Close to this point is the ear with the tympanum (the tympanic bone of mammals derives from the angular bone of the reptilian mandible) and the stirrup (stirpes). In mammals the quadrate and articular bones have become the anvil (incus) and the hammer (malleus) which, together with the stirrup, constitute the chain of ossicles of the middle ear. The chain of ossicles represents the only characteristic that can truly be described as exclusive to mammals. The function of these tiny bones is to perfect the transmission of sound from the middle ear membrane (tympanum) to the inner ear. In the more primitive animals, as in amphibians, reptiles and birds, the tympanum is an external organ, either on the surface or close to it. Vibrations are

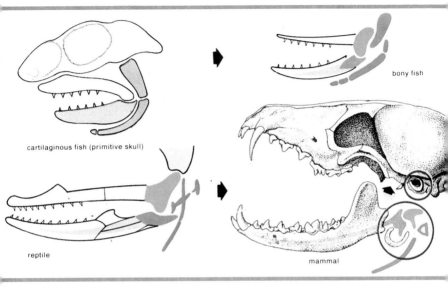

▲ Evolution of the maxillary arch and the formation of the ear ossicles

transmitted from the tympanum to an inner membrane (bounding the inner ear where the vibrations become nerve impulses that are sent to the brain) by means of a single ossicle, the stirrup, which links these two membranes.

The stirrup developed from the hyomandibular bone, and the tube containing it is derived from the spiracle or branchial orifice (breathing hole) of selachians and primitive fishes. In mammals, however, the external tympanum slopes at the bottom at the point of articulation with the mandible and is linked to the auditory meatus, or channel, of the external ear. Between these two membranes, there is the chain formed by the stirrup, the anvil and the hammer, articulating with one another. The function of these bones is to amplify faint vibrations and soften those that are too loud. Situated between the middle ear and the mouth cavity is the Eustachian tube, the function of which is to maintain equal pressure on either side of the tympanum. Other important features of the mammalian skull are the articulation of the cranium and the first cervical vertebra by means of two occipital condyles; the formation of the secondary palate; and the enlargement of the cranium to allow for the increased volume of the brain.

In primitive forms of mammal the temporal fossa, which is large and near the surface, is bounded below by the zygomatic arch (comprised of the jugular bone and the zygomatic process of the squamous bone) which is joined by the orbital fossa. In specialised forms there is a rear wall to the orbit (eye socket) which may be almost complete, as in the human skull.

The skull consists basically of two parts, a neurocranium, which is the case containing the brain, and a splanchnocranium, which is the facial part of the skeleton with the jaws, eyes and nose. The teeth are carried in the upper jaw by the maxillary and premaxillary bones and in the lower jaw, or mandible, by the dentary bone. It has already been mentioned that the majority of mammals have hetorodontic teeth and also have two sets of teeth (diphyodont), during their life. The first set consists of milk or deciduous teeth which are later replaced by a permanent set. The front teeth are the incisors, which are usually flat and fairly small, although in the case of rodents the incisors are much larger and of continuous growth. Next come the canines, which are sometimes very large as in the carnivores and wild boars, but sometimes absent in rodents. Behind these are the premolars and molars. These always have roots and vary in shape. In carnivores, for instance, they are sharp and cutting, while in herbivores, ungulates and rodents, they are flat and suitable for grinding. Those of some rodents grow continuously to replace surface wear. The form of the teeth provides a fairly reliable indication of the diet and life style of the mammal concerned. The number of teeth varies according to the animal and primitive species such as the marsupials and insectivores usually have more teeth than other mammals. Since teeth are the most easily preserved parts of a fossil skeleton, they are important in classification.

The simplest and most methodical way of describing a mammal's tooth pattern is to apply what is called the dental formula. This lists, in order, the number of incisors, canines, premolars and molars in each jaw, from front to back. Each type of tooth is indicated separately, those of the upper jaw above a fraction sign, those of the lower jaw below. For the sake of simplicity, the figures relate to one half of the jaw only, so that to arrive at the total one has to double the number given in the formula. The following, for example, is the dental formula of a hare: $I\frac{2}{1}$, $C\frac{0}{0}$, $PM\frac{3}{2}$, $M\frac{3}{3}$. That of a horse is $I\frac{3}{3}$, $C\frac{1}{1}$, $PM\frac{3}{3}$, $M\frac{3}{3}$; and that of a brown bear $I\frac{3}{3}$, $C\frac{1}{1}$, $PM\frac{3-4}{2-4}$, $M\frac{2}{3}$. These formulae are further simplified, respectively, as follows: $\frac{2.0.3.3}{1.0.2.3}$, $\frac{3.1.3.3}{3.1.3.3}$, $\frac{3.1.3-4.2}{3.1.2-4.3}$. Thus the brown bear has a total of 36–42 teeth, depending upon the number of premolars. The dental formula of man is $\frac{2.1.2.3}{2.1.2.3}$, giving a total (16×2) of 32 teeth.

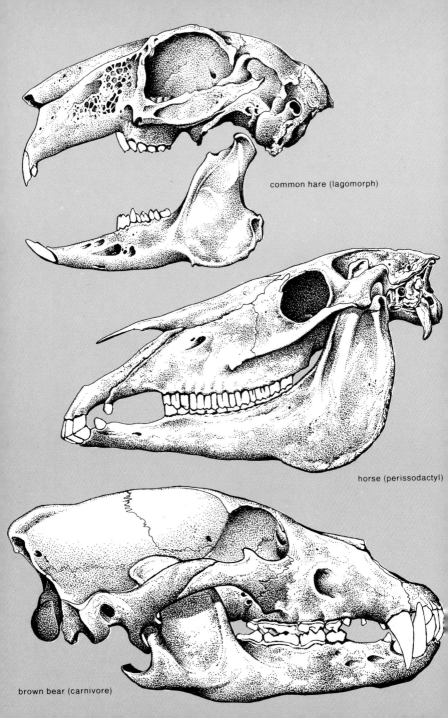

common hare (lagomorph)

horse (perissodactyl)

brown bear (carnivore)

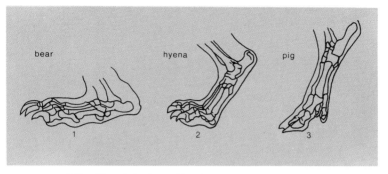

▲ The different structures of feet in
(1) plantigrades; (2) digitigrades; (3) unguligrades

Form and movement

The body shape of mammals is extremely varied and closely
linked to their life habits. For example, although leg and foot
structure is basically of a standard type, the number and shape of
the individual bones that make up the limbs have undergone considerable modification according to the way in which a particular
animal moves.

The following section looks at some of the distinguishing
features of the axial skeleton (spine, ribs and sternum) and of the
appendicular skeleton, consisting of the limbs and girdles that join
the appendages to the axial skeleton (the shoulder or pectoral
girdle in the case of the forelimbs, the pelvic girdle in the case of
the hind limbs). The shoulder girdle may or may not contain a
clavicle or collarbone. It is present in all the more primitive forms
of animal, permitting full freedom of movement in all directions to
the forelimbs. It is, however, reduced or absent in all running
mammals who, because of their environment, need to be able to
perform broader movements forwards and backwards.

The majority of mammals are land creatures. These include
so-called plantigrade, digitigrade and unguligrade species,
distinguished by the manner in which they place their feet on the
ground, as shown in the diagrams above. In plantigrades such as
the bear the entire surface of palm and sole is in contact with the
ground. This is the most primitive type of gait, similar to that of
the synapsid reptiles and primitive insectivores which support
themselves on five-toed feet. It is used by a very large number of
land mammals including man. Almost all the carnivores are
digitigrades, walking on their toes. This effectively increases the
length of their limbs, the weight of the body being supported by

▲ Tracks of a bear

▲Alpine hare (*Lepus timidus*)

the tips of the digits, without the wrist and heel having to make direct contact with the ground. The unguligrades include the swift-running herbivores and as a result of their movement on the tips of their toes, claws have developed into hooves. This form of locomotion increases the effective length of the limbs to an even greater extent. The bones of all four limbs are not only elongated but also fewer in number. Virtually only the third and fourth toes remain in the even-toed ungulates (Artiodactyla). The metatarsals and metacarpals of these toes are usually fused so as to form the cannonbone, and the phalanges have developed into the bifid hoof. The second and fifth toes (especially in the Bovidae) are reduced. A more specialised situation has arisen in the case of the odd-toed Perissodactyla, such as the horse, where the cannonbone is formed by the third metapodial alone, only the third toe coming into contact with the ground. The hoof is single, the other toes have disappeared and the second and fourth metapodials are reduced to bony plates.

There are some mammals which adopt forms of locomotion that cannot be strictly categorised since they do not always use a single type of gait. By and large, however, small mammals with

31

▲ Cheetah (*Acinonyx jubatus*)

short legs tend to be plantigrades and large mammals, such as elephants, and rhinoceroses, digitigrades or semi-digitigrades. Apart from the manner in which they rest their feet on the ground, mammals move in a very different way to the reptiles from which they are descended. The latter rest the belly on the ground between one movement and the next. Mammals, with their more specialised limbs, are able to move with their bodies kept permanently above the ground. They rest either by lying on one side or in a seated position. They have completely abandoned the awkward, serpentine movement employed by reptiles, and are able to move in a straight line. Only the seals, adapted to living in water, move about reptile-fashion when on dry land.

All slow-moving mammals walk by lifting one foot at a time. At greater speed the movements are more rapid and complex. Most mammals break into a trot and at higher speeds into a gallop. However, some amble rather than trot, and instead of galloping perform a series of leaps. Trotting is the normal gait of many carnivores and of ungulates. It is a symmetrical form of locomotion, comprising movements in which the body is supported by a foreleg and hind leg on opposite sides, alternating with brief intervals when the body is not supported at all.

▲ Roe deer (*Capreolus capreolus*)

The amble, on the other hand, which is the characteristic gait of camels, consists of moving both legs on the same side forward, while the legs on the opposite side are moved backwards. Certain mammals neither trot nor amble but use various types of hurried steps or short gallops. Even the true gallop has many variations, but basically it is an asymmetrical gait, employed by all the faster-moving mammals.

Some mammal species proceed by leaping (a symmetrical movement) or bounding, as is the case with many carnivores, and lagomorphs. Certain deer and antelopes use a strange kind of gallop in which they jump off the ground with all four legs simultaneously. A few mammals adopt a bipedal or two-legged gait but man is the only mammal who is able to walk normally in this fashion. Some monkeys occasionally walk on two legs on the ground but generally use their arms as well. All other bipedal mammals leap with their hind legs, using the tail as a balancing organ, as is the case with kangaroos, kangaroo rats, jerboas, gerbils and jumping hares. Some mammals can move at astonishing speed, although only over short distances. The cheetah has been known to reach 113 km per hour and certain antelopes can do more than 95 km per hour.

There are many mammals, however, which are specially adapted to move in their particular environment. Tree-dwelling mammals often possess special structures such as prehensile tails, opposable fingers and, in the case of gibbons and sloths, highly developed claws. Suckers on fingers and toes are a feature of a few specialised arboreal mammals, while others are experts in gliding flight, such as the flying squirrels and phalangers. Bats, however, are the only mammals specialised both in structure and behaviour for true flight.

The other specialised form of locomotion is swimming. Almost all mammals are capable of this, although some such as man and the anthropoid apes have had to learn to overcome an innate fear before venturing into water. Depending on the animal and its environment mammals have adapted to aquatic life to varying degrees. For instance, otters.are equally at home on land and in water and their specialisation is fairly limited. Seals, on the other hand, have more pronounced adaptations in the form of flippers and a very streamlined body. Those mammals that live permanently in water, such as manatees, dugongs and whales, display all manner of specialisations.

External structure

External body features, for example the skin and its derivatives (hairs, claws, horns, glands, etc) represent a boundary zone between an organism and its surroundings. These features are, therefore, closely associated with an animal's way of life in its particular environment. The entire central nervous system is devloped from the embryonic layer of skin.

The actual structure of the skin comprises a thin outer layer, the epidermis, which covers the thicker, tougher true skin or dermis. The epidermis is in turn continually producing a horny layer (stratum corneum) of dead cells that prevents the skin drying out. The dermis is provided with a mass of sensitive nerve endings that react to a variety of stimuli including changes of temperature, touch, pressure and pain. The dermis is also provided with a dense network of blood vessels supplying nutrition to the skin surface. The skin also has sweat glands, which are vital for regulating body temperature, and sebaceous glands that produce fatty secretions for lubricating the hairs.

The thickness of the coat and quantity of hairs vary considerably according to the species and its lifestyle. The majority of mammals have some degree of coat. However, as a result of adaptation to their watery environment the cetaceans and sirenians have only a sparse covering or completely lack any coat. The hairs develop from the thickening horny part of the

▲ Humpback whale (*Megaptera novaeangliae*)

epidermis and extend down into the dermis where they form tiny papillae linked by a network of blood capillaries that supply moisture. The soft epidermic cells above each papilla multiply, gradually acquiring more keratin as they push upwards, eventually forming a fibrous shaft. The hair thus formed consists of a cylindrical, filamentous shaft which appears above the skin surface and of a root implanted at an angle in the epidermis. The structure of each hair is quite complex, comprising a central medulla, often with cavities; a cortical layer (Huxley's layer) containing the pigment granules that determine colour; and a protective outer cuticle. The form of the hair depends on the proportionate sizes of the medulla and the sheath, the shape of the scales composing the cuticle, and the structure of the follicle or recess containing the hair. There are two basic types of hairs: the long, stiff covering ones known as contour hairs, and the thinner, denser ones underneath known as woolly hairs. The appearance of the coat depends on which of these two types of hair predominates. The woolly hairs create an insulating thermal layer and are, therefore, more numerous in young animals and in the winter coats of mammals that live in temperate and cold regions. In domestic breeds of sheep the contour hairs have been eliminated by artificial selection, and there has been a great increase in the number of woolly hairs which constitute the under-fur or fleece. This wool can be spun because of the structure of scales in the surface layer which enables the hairs to hook on to each other, so forming a felt.

The contour hairs develop from deeper hair follicles and the groups of secondary woolly hair are arranged around them. This helps in the processing of certain types of fur, such as the pelts of

▼ Crested porcupine (*Hystrix cristata*)

the beaver and coypu. In both species the long, bristly contour hairs can be removed, leaving only the very soft woolly hairs that give the finished beaver skin its velvety feel and appearance.

Characteristic wrinkling of the hair depends on the oblique position and shape of the cutaneous follicles and the resultant angle of growth of the hairs. If the follicle is curved rather than straight, the hairs will be curly or wavy. The coat is also subject to periodic moult and may change length and colour according to the animal's sex, age, and the season. Mammals do not usually have a brightly coloured coat. This is probably related to the fact that most mammals do not distinguish colours, although certain primates display patches of bright colour in their coats. The hairs making up the coat basically contain dark pigments (melanin), producing various shades and patterns of brown, black, grey and white.

Other special cutaneous structures developing from the hairs include vibrissae, stiff tactile bristles connected to more specialised nerve endings that are generally situated on the muzzle (the whiskers, which are highly developed in all nocturnal mammals). These may also be found in other parts of the head or body. In some mammals the hairs have become so stiff and strong that they form spines, which are often packed tightly together. Porcupines and hedgehogs have either simple spines or very stiff quills which are embedded in quite a complicated manner. The spines of the crested porcupine (*Hystrix cristata*), for example, break off easily at the base. The tail is furnished with special hollow quills which, when the animal moves its tail in a threatening fashion, produce a surprisingly loud noise. Other mammals with spines are the echidnas.

Another form of protective covering is provided by scales. Those of pangolins (illustrated on p. 39), have virtually the same origin and structure as ordinary hairs, whereas the armour plating of armadillos consists of small bony scutes in the dermis, covered by a horny layer of epidermis.

Additional specialised structures deriving from the skin are nails, which serve to protect the tips of the fingers and toes and can vary in form. The only mammals without nails are cetaceans (whales and dolphins), sirenians (manatees and dugongs) and some otters. The nails are formed from a very thick layer of keratinised dead cells, the bases of which grow from roots containing many blood vessels. The most primitive type of nail is the claw. The nails of primates and man are reduced forms of claw. The typical hoof of ungulates, covering the sides and upper part of the toe as well as part of the lower surface, is a refined development.

▼ Longitudinal section of sweat gland and hair follicle of bat

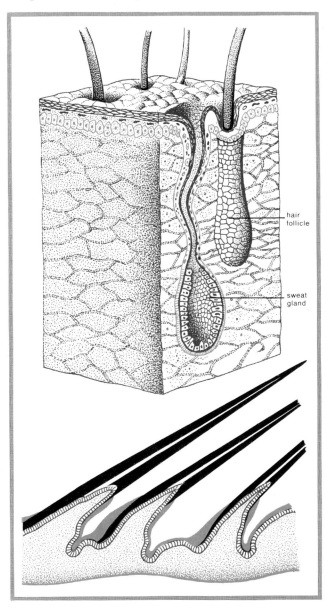

hair
follicle

sweat
gland

▲ Longitudinal section of epidermic scales of a pangolin

Ungulates are also unique among mammals in possessing horns. These are partly epidermic in origin but have a central support formed from a bony outgrowth of the skull's frontal bone. There are two main types of horn, those of the Cervidae and those of the Bovidae. The former, known as antlers, are solid and are shed at regular intervals; the latter are hollow and permanent. Both forms only begin to grow after the animal is born. Among deer antlers are normally found only in the male, but they are present in both sexes in the case of the reindeer and the caribou (*Rangifer tarandus*). In certain species of small deer, such as the chevrotains (*Tragulus*), the musk deer (*Moschus moschiferus*) and the Chinese water deer (*Hydropotes inermis*), there are no antlers in either sex. In these cases the upper canines, especially of the males, are overdeveloped so as to form long tusks, which are used instead of antlers as defensive weapons.

▼ Scales of small-scaled tree pangolin (*Manis tricuspis*)

▲ Solid antlers of roe deer in velvet (*Capreolus capreolus*)

▲ Hollow horn of chamois (*Rupricapra rupricapra*)

The antlers of deer originate as small bony knobs on the frontal bones of the forehead, covered with skin and hair, constituting the beams. During the period of growth they remain completely covered in skin, with short, compact, velvety hairs. This covering is called velvet. When the growing period is over the blood vessels at the base of the antlers (the pearl) close, interrupting the supply of blood to the velvet. This causes the skin covering to dry up and die, after which it breaks off. At this stage the animal itself helps in the shedding of the dead velvet by rubbing its antlers against trees and shrubs. All that remains is the bony core, which it sometimes rough, sometimes smooth. After a certain period, which varies according to species, the bony material is also resorbed at the point where the pearl and the beam join, causing the base to weaken and the antlers to be shed.

Soon afterwards, the cutaneous tissue of the beam begins to grow again, covering the bare bony knob and initiating the formation of a new antler. The shedding and regrowing process generally occurs once a year, and every year sees an increase in the number of branches or tines until the animal has attained the maximum number appropriate to its species. After deer reach a certain age the size of the antlers and the number of tines tend to decrease. The shape and number of tines varies greatly depending upon the species concerned, ranging from the small, barely forked antlers of the roe deer (*Capreolus capreolus*) to the magnificent, complex growths of the European red deer (*Cervus elaphus*) and the wapiti (*Cervus elaphus canadensis*). Sometimes they appear in the form of flattened plates (palmate). This is as a result of the fusion of the beam and the lateral branches. The elk (*Alces alces*) and fallow deer (*Dama dama*) both have palmate antlers.

It is hard to explain the advantages of this complicated procedure of continuous regrowth and shedding of antlers. It involves enormous expenditure of energy, particularly for the larger deer. Within a few months they have to consume up to 50 kg of calcium in their diet which consists exclusively of vegetation. Centuries ago there were even larger forms such as the huge Irish elk (*Megaceros giganteus*), with a span of enormous palmate antlers up to $3\frac{1}{2}$m wide. There was also a giant deer (*Cervus dicranius*) during the Pleistocene epoch, with a span of about 2 m. The development of branched antlers occurred relatively late in the evolution of the deer, and it seems that it was not until the Pliocene that deer with ramified antlers appeared.

▼ Diagrams showing how a deer sheds its antlers (1–2) the broken bone is replaced by connective tissue; (3–4) the growing skin causes shedding

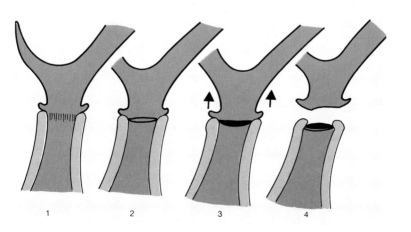

1   2   3   4

Some even had three antlers and there was one species with a single antler. All were descended from older species which either had no antlers or were simply endowed with small bony nasal and front knobs.

Among the Bovidae, all males carry horns. In many species the females also have them although the latter are often smaller and thinner. No male of any wild species is without horns, but this is not the case in some domesticated breeds. Certain species of sheep carry either four or six horns and the Indian four-horned antelope (*Tetracerus quadricornis*) is so named after that uncommon feature.

In Bovidae the bony cores are outgrowths of the frontal bones and are very often hollow because the cavities of the frontal air sinuses extend inside them. They are covered with a horny sheath formed by keratinised epidermic cells that multiply continuously at the base. The shape of the horns is extremely

▼ Fight between fallow deer stags (*Dama dama*)

variable. They may be simple and straight, as in the different species of oryx; curved to a lesser or greater degree, forwards, backwards or even in spirals, as in sheep, antelopes and goats; hooked, as in the chamois (*Rupicapra rupicapra*); or with huge frontal bosses, as in the African buffalo (*Syncerus caffer*), the gnus (*Connochaetes*) and the musk-ox (*Ovibos moschatus*). The shape depends on different rates of growth during their development, but the process is fairly complicated. The horns of many sheep, antelopes and goats have rings resulting from variations in growth rates. In some species, such as the Rocky Mountain goat (*Oreamnos americanus*), the chamois and the mouflon (*Ovis musimon*), the number of rings is an indication of the animal's age, since horn growth is arrested during the winter.

The pronghorn antelope (*Antilocapra antilocapra*) of North America has rather unusual horns. The bony cores of these branched horns are covered with horny sheaths which are shed annually. After the breeding season new sheaths are formed and in due course the horns are again fully developed.

In giraffes the horns, carried by both sexes, are bony projections covered with skin and velvet, similar to the antlers of deer, but permanent. The same type of horn, though present only in males, is found in the okapi (*Okapia johnstoni*).

Mention has already been made of the sweat glands and sebaceous glands of the skin. From these are derived a number of other specialised cutaneous glands found in many mammals. Thus the sticky reddish secretion of the hippopotamus, which helps to protect the skin when it is dry and to prevent sunburn, is the product of modified sweat glands. Other characteristic glands are modified versions of sweat glands and sebaceous glands. Among the latter are the olfactory or scent glands, present in most mammals. The odours produced by these glands are generally released in the form of oily or fatty secretions. They have a variety of functions that are extremely important to mammals as they rely primarily on their sense of smell. Among gregarious species they help to keep groups of animals together (as, for example, in sheep, which leave traces of an odour produced by glands situated between the toes). Alternatively, scent may be used in order to signal an individual animal's presence to a potential mate. It may also serve to mark out territory and so keep away intruders.

Olfactory glands may be situated all over the body. They are, for instance, found behind the ear of chamois, on the neck of camels, on the forehead of elephants and on the flanks of shrews. Hyraxes and peccaries have such glands on the back. In rabbits scent glands are situated below the chin, and in deer and antelopes under the eye. Very often, however, scent glands are situated in

43

▲ Female yellow baboon (*Papio cynocephalus*) with young

the anal or perineal region. Anal glands produce odorous secretions that are usually unpleasant. They generally serve to demarcate territory but sometimes they are connected with sexual activity, as with the hyena. They can also be deliberately used as a means of defence, as with the opossums and the skunks of North America. The stench produced by the secretions ejected from the anal region of skunks is singularly strong and unpleasant. Among civets, the secretions of these glands, which are especially large, collect in a pouch-like structure and are used for manufacturing perfumes. Other olfactory glands are associated with the excretion of urine. All carnivores habitually mark their territory with urine, but in many mammals the odour is heightened by secretions from specialised preputial glands.

By far the most important glandular structures of the skin, however, are the mammary glands. These are exclusive features of the class to which they lend the name. They are modified forms of sweat glands and sebaceous glands. Situated in the chest and

▲ Lioness (*Panthera leo*) suckling cubs

abdominal regions, they appear at an early stage in the embryos of mammals. The glands, although present in both sexes, are functional only in the female. They are located in the mammae or breasts, their positions, number and development varying according to class. After the first birth, they become active and under the stimulation of a special hormone, prolactin, they produce secretion of milk, which provides nourishment for all baby mammals until they are old enough to feed themselves.

Internal structure
Having briefly described the tegumentary apparatus (covering structures) and the skeletal system (supporting structures) of mammals, the following section deals with the more important internal mechanisms of the body. Attached to the framework of bones are the muscles which cause the bones to perform movements in response to stimuli received through the nerves. In addition to the muscles linked to the skeleton there are other

Digestive apparatus ▶
of ruminant:
  (1) rumen;
  (2) reticulum;
  (3) omasum;
  (4) abomasum

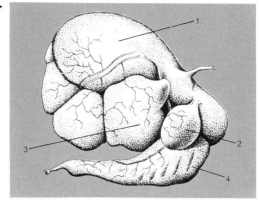

involuntary muscles incorporated in the various organs of other systems – digestive, respiratory, circulatory, excretory, nervous and reproductive. Whereas the nervous system represents the central control of the body and the reproductive system ensures its continuation, the other four major systems of the body function as its motor.

These systems are situated inside the visceral body cavity. This is divided into two parts by a muscular diaphragm, which is exclusive to mammals and extremely important for breathing. The front or thoracic part of the diaphragm contains the heart and lungs, and the rear or abdominal part contains the other viscera or intestines. Heart, lungs and intestines are organs that undergo continual changes in size, depending on their function. They are each wrapped in double-walled sacs (the pericardium, pleura and peritonium respectively) that make such modifications easier. The organs of the digestive system are also attached to the posteria wall of the abdomen by the mesentery.

The basic structure and fundamental functions of these body systems are fairly similar to those of the human body, but in many mammals they are modified for special purposes. This is particularly so in the case of the digestive system, which varies according to the animal's type of diet. In ruminants, for example, the stomach is subdivided into four sections, each with different functions. The first of these separate compartments is the rumen. This swarms with bacteria and protozoa capable of breaking down the cellulose contained in vegetable matter, so that the food, having been rapidly swallowed, can pass into the second stomach cavity and be regurgitated. It is then chewed at leisure (ruminated)

◀ Leopard (*Panthera pardus*)

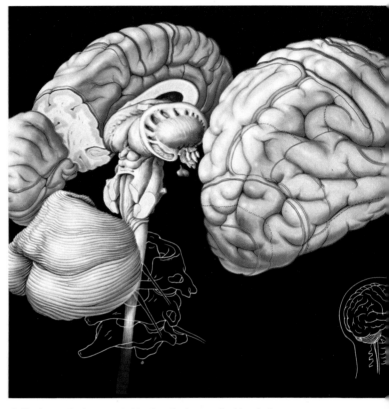

▲ The human brain separated to show the two cerebral hemispheres (cortex), the cerebellum and the spinal cord (see pp. 51–2)

and despatched to the other compartments. Among herbivorous mammals that do not possess a rumen, the cellulose can be digested with the aid of bacteria present in the caecum or blind gut of the intestine, which is a highly developed appendage. The digested matter is first absorbed by the intestines, then circulates around the body where it is finally distributed among the various organs and tissues.

The blood circulation system is a wholly enclosed, two-way process, and the heart (as in birds and some reptiles) has two auricles and two ventricles. Closely linked to the circulation system is the respiratory apparatus, in which the lungs bring

▲ Rats have a keen sense of smell
which is aided by long, sensitive whiskers

about gaseous exchanges. The substance which briefly unites with
carbon dioxide and oxygen, conveying oxygen to the tissues, is
haemoglobin. This is the colouring matter contained in the
erythrocytes or red corpuscles which, in mammals, lack a
nucleus. The excretory system, also closely related to the blood
circulation, is the means of eliminating the toxic nitrogenous
substances produced by the digestion of proteins. The bile-
secreting liver transforms these toxic nitrogenous substances and
the finished product, called urea, is carried in the bloodstream.
The urea is finally filtered by the kidneys, to be excreted with the
urine.

▲ Philippines tarsier (*Tarsius syrichta*)

The nervous system

In all animals the nervous system is the structure which directs and co-ordinates all the body's activities. The central nervous system consists of the brain, enclosed and protected by the bony case known as the cranium, and the spinal cord, which is contained in the vertebral canal, a rigid, articulated tube. The peripheral nervous system is made up of all the nerves, and links the central system to every part of the body. It carries messages to the effector organs and tissues such as glands and muscles and receives information from the sensory organs. Intestinal

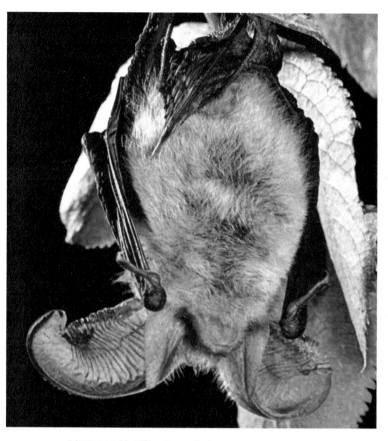

▲ Long-eared bat (*Plecotus auritus*)

activity, in particular, is controlled by the sympathetic and parasympathetic system, made up of chains of ganglia on either side of the spine. This is linked to the internal organs and to the central system. In mammals the nervous system is extremely complex, the brain in particular being composed of an enormous number of inter-linked cells. The mammals are, therefore, distinguished from other vertebrates by the advanced development of the cerebral cortex of the brain.

The volume and complexity of the cortex increase in the more highly evolved species. Thus the cerebral hemispheres of placental mammals are larger and richer in the bands of nerve tissues known as commissures than in the monotremes and marsupials. These two groups lack the so-called corpus callosum or transverse commissure connecting the hemispheres. The bodies of the nerve cells situated close to the surface of the cortex constitute the brain's grey matter, and the underlying fibrous substance of these cells is the white matter. It is obvious that the larger the area of the cortex, the greater is the number of cells. In the more complex brains the area is increased by the folds that make up the convolutions, thus making possible a wider range of responses to different stimuli. In the most primitive mammals a large part of the brain consists of olfactory bulbs. These relate to the sense of smell which in their case is extremely important. Smell, however, is of little significance to cetaceans which live in the water, and they, like man, depend mainly on their senses of sight and hearing. There is no space here to describe the structure and function of the eye. However, it is worth mentioning that in the majority of mammals the retina is composed chiefly of rods rather than cones. This makes it possible for them to note and swiftly appraise any kind of movement but not to distinguish colours. Most mammals have probably lost this ability to see in colour, for it seems to have been an endowment of their reptile ancestors and may have been reacquired, possibly as a secondary function, in certain groups such as primates, cats and giraffes.

Many mammals employ ultrasonic frequencies for purposes of communication. Recent research has revealed that whales and bats have complex communication patterns.

The reproductive system
The reproductive method of the majority of mammals (viviparity or live birth), is highly complex. In viviparous animals the embryo develops inside the uterus of the mother. It receives nutrition by various means and for a gestation period of varying length depending on the species. The newly born offspring will have reached a certain stage of development according to the length of gestation. This is one of the fundamental characteristics of mammals. As far as the reproductive mechanism is concerned, there are always two sexes and fertilisation occurs internally. The reproductive organs consist basically of a pair of gonads – the female ovaries and the male testes – which connect with ducts (the oviducts or Fallopian tubes in the female and the vas deferens in the male) that carry the ova and spermatozoa to the respective organs of copulation. The mammalian egg, which is comparatively small, travels through the oviduct into the uterus, the stem end of which leads to the vagina. This, in turn, opens to

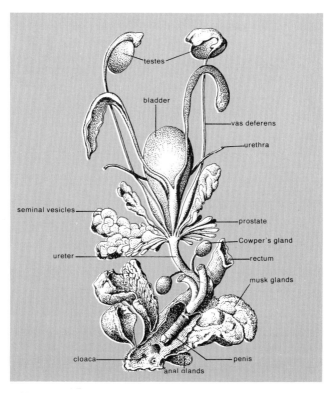

▲ Urinogenital system and scent glands of the male beaver (*Castor fiber*)

the exterior as part of the urinogenital mechanism. Fertilisation generally takes place in the oviducts when an egg unites with a spermatozoon contained in the seminal fluid. The seminal fluid is ejaculated into the vagina by the male penis. The fertilised ovum attaches itself to the wall of the uterus and the embryo then develops. In the more highly evolved eutherians the embryo is in direct contact with the mother by virtue of the placenta. This structure develops from the thickening of the egg membrane or chorion and is affixed to the wall of the uterus.

The gestation period varies: in mice it is about twenty days, in dogs two months and in the sperm whale up to sixteen months. The period is generally shorter among marsupials, the young being born at a very early stage of development because, in the absence of a true placenta, they have to be fed on mother's milk.

How Mammals Live

Having briefly looked at the definition and structure of mammals it is interesting to see how they live, and examine some of the more elementary features of the field of biology known as ethology. This is the study of animal behaviour, or in simple terms, what they do and why they do it.

Many animals exhibit behaviour that to a large extent conforms to a set pattern and is predictable. This is called stereotyped behaviour, and is conditioned by responses to particular stimuli. Stimuli can be produced from a number of sources, such as individuals of the same species, by other animals or by surrounding objects. Mammals in particular, with a very complex brain made up of an enormous number of nerve cells linked by connective tissue and commissures, display an extremely varied range of responses to such stimuli and this enables them to survive in all manner of different situations. This type of behaviour results from the combination of several factors. Some behaviour patterns are derived from information inherited from the parents through the genetic code; some are learned by imitating the behaviour of the parents and others depend on learning by a trial and error process. When the solution to a particular problem is found by introspection, using a mental process, logic, which merges memories of a past experience with the reality of the present situation, one can describe it as intelligence.

Among the most important types of behaviour affecting the life and survival of an animal are those directed towards self-defence and protection from dangers. There are several alternatives for an animal facing perils, whether from the environment, from other

animals or from creatures of its own kind. It can adopt a system either of passive or active defence. The main weapons available for passive defence are the appendages of the skin covering, such as spines, scales and bony scutes. These enable an animal to curl up, like the hedgehog, into a spiny ball so that it cannot be attacked by an ordinary predator. Equally effectively, some, such as the armadillos, can roll into a ball of tough, bony, discontinuous armour-plating. Some animals, such as the skunks, protect themselves by ejecting a foul-smelling fluid from specialised scent glands. Other mammals escape predators by staying quite motionless, hidden by their cryptic coloration (a form of camouflage). The American opossum (*Didelphis marsupialis*), if attacked and unable to flee, feigns death – hence the expression 'playing possum'.

Active defence entails the use of weapons of an offensive nature. Almost all mammals use teeth and claws (structures principally used for other purposes) in such circumstances. The teeth are mainly intended to obtain food and to chew it, and the claws are mainly intended to protect the tips of hands and feet. Both may be modified in various ways so as to become effective weapons of defence or offence. The teeth may also sometimes be specialised, as is the case with the highly developed canines of various species of wild pigs.

In addition to these characteristic weapons there are others that are more specialised. The male platypus (*Ornithorhynchus anatinus*) is equipped with poisonous spurs on the inside of its hind feet, yet their precise function is not known. Among the perissodactyls, the rhinoceros has a horn on the forehead. Although its primary function is to stimulate its partner prior to mating it is also used as a weapon of defence. The development of the horn of the artiodactyls (Cervidae and Bovidae) is apparently related to the reduction in size of the front teeth, especially the incisors and canines. In some deer, as has been said, upper canines have been transformed into strong tusks to compensate for absence of antlers.

The offensive weapons of mammals, whether or not specialised, are either used for protection against predators or by

▲ Hedgehog (*Erinaceus europaeus*)

the predators themselves to capture prey. Therefore, their function is to ward off dangers and to procure food. In the case of fights that occur within the same species (intra-specific combats), these weapons will not be used seriously but merely as exhibitions of strength designed to force the retreat of a weaker rival without a real fight. Combat behaviour in mammals is generally intended to avoid actual fighting so that the combat itself is to a great extent ritualised and rarely ends in death. The only situations in which death or severe injury may occur are those in which the number of

Three-banded armadillo ▶
(*Tolypeutes*) rolled up into a ball

▲ Armadillo (family Dasypodidae)

individuals has exceeded the number that can be supported by the resources of the environment. In such cases of overpopulation, or in captivity when there are artificial boundaries to the animal's habitat, there can be increased aggressiveness in some individuals. Under these circumstances other individuals of the group are also prevented from fleeing the aggressors themselves. All these situations are abnormal in the sense that the habitat is not adapted to the needs of the animals occupying it. The result is invariably an alteration in normal behaviour patterns.

57

Combats among individuals of the same species are always closely associated with territorial control and reproduction. Virtually all mammals exhibit aggressive behaviour during the mating season, as is seen in ungulates, seals and felines. Among ungulates, for example, classic and familiar cases are the combats staged by deer. For the greater part of the year these animals live in segregated groups of males and females. However, in the rutting season they exhibit territorial behaviour, the males staking out territory and the females forming groups within that territory. This is the time when fights break out between rival males. One male will announce his claim to a piece of territory by emitting special sounds (belling) which serve to keep other males at a distance. Should an intruder dispute possession of another male's territory, a very violent and noisy fight will ensue. However, these seldom result in injury on either side. The loser – almost always

▼ Grant's gazelles (*Gazella granti*)

the challenger – is not even followed by the victor. Similar combats, perhaps even more ritualised, occur among seals, who are also polygamous animals which defend their own harems of females against other males. The fighting of domestic cats on heat is a familiar experience, and the same sort of behaviour is found among other wild members of the cat tribe. Tigers, for example, are solitary creatures which come together only in the mating season. Attracted by the characteristic call of a female on heat, several males will fight fiercely to possess her. Here too, animals that are extremely powerful and furnished with dangerous weapons such as long, pointed canines and sharp, well-developed claws, stage combats that are no more than ritual exhibitions of individual strength. As a result of this behaviour it is generally the larger and stronger males who leave descendants since it is they who have the opportunity to mate, even though other less

aggressive males very often take advantage of combats to mate with temporarily unoccupied females.

Among gregarious animals ritual combat is markedly specialised. One of the best known examples, comprehensively studied by ethologists in recent years, is that of the wolf. The pack has a social structure in which every individual finds its own determined level in the hierarchy, with an order of precedence that is accepted by the community. This system of social hierarchy is a common phenomenon among mammals and also exists in birds (pecking order). The dominant members of the wolf pack impose their will upon subordinates by means of aggressive expressions, such as growling and snapping of bared fangs. Should the subordinate animals offer a show of resistance they may fight. When the subordinates eventually yield, they assume the typical attitudes of cubs, rolling over on their back and exposing the most vulnerable parts of the body to their superiors. This gesture of submission inhibits the aggressiveness of the dominant animal, as happens in the relationship between father and cubs. There are other ritual expressions of the face and movements of the tail which are used as signals of friendship, submission and sociability. These attitudes can be recognised in the domestic dog.

Stereotyped behaviour such as this is to be found among most mammals, but it is particularly evident in those with a more complex degree of social organisation. It may also serve the purpose of maintaining links among individuals of the same species and isolating themselves from other species. Sometimes there is a notable outburst of aggressive behaviour among different groups of a single species, as is the case with the common brown or sewer rat (*Rattus norvegicus*). In the wild each population has its characteristic scent and will show aggressive behaviour towards rats belonging to other communities with different odours. In captivity an individual belonging to another clan will immediately be attacked and killed.

Apart from aggressiveness, and various other forms of defensive behaviour that have already been mentioned, the majority of mammals seek safety in flight. In many species there is a precise 'flight distance', as can easily be recognised in herbivores living in open, exposed habitats. When confronted by danger at close quarters, most animals remain calm and motionless until the enemy gets to a certain point, and then they take to their heels. If there is no time to get away, a fight will ensue and the animal will defend itself as best it can. Some mammals, however, if surprised by a predator within the critical distance will stay still and surrender without a struggle. This happens with hares attacked by stoats or some of the larger African herbivores

▲ Male lions (*Panthera leo*) in combat

assailed by packs of hunting dogs. A special case, already
mentioned, is the feigned death behaviour of the opossum.

Many mammals take refuge in burrows, shelters or nests.
These various types of shelter are proper homes where an animal
will be protected from extremes of climate and from predators.
This is the base from which it will venture into surrounding
territory in search of food and to which it will eventually return. It
is here, too, that the young are born and reared. Whereas the nests
of birds are temporary structures for brooding and raising chicks,
those of mammals are stronger and more durable, sometimes
being used for a number of generations. Among the mammals
that build nests in vegetation are rice rats (*Oryzomys*) which
construct elaborate spherical nests suspended between tall grassy
stems, as well as dormice and squirrels whose nests are situated in
trees or in bushes.

▲ Alpine marmot (*Marmota marmota*)

The majority of rodents, however, excavate underground burrows, like those of the prairie dogs (*Cynomys*) and the viscacha (*Lagostomus maximus*) of the South American pampas. These burrows contain numerous galleries of varying depth. Other types of burrow, such as those of fieldmice, are much simpler and shallower, with nests often situated just below the turf or sometimes under stones. Rabbits also dig tunnels but do not usually build real nests except when giving birth. Hares make use of shelters above ground, somewhat similar to the simple depressions of some ruminants, deer and antelopes. Most large ruminants, however, do not have any kind of permanent shelter.

The insectivores behave more or less like the smaller rodents. Moles build a complicated nest consisting of a chamber fairly near the surface, linked to others by a network of tunnels that are habitually used in their life below ground. Badgers also dig lairs, although some other carnivores such as foxes take over holes already in existence.

The same sort of behaviour is to be found among the marsupials, but in their case protection of the young generally does not entail the use of a nest, since the marsupium or pouch is itself a kind of portable nest. Kangaroos do not have any form of

▲ Red fox cub (*Vulpes vulpes*) at entrance to lair

permanent habitation, for like the big ungulantes, they lead a wandering life. Other marsupials, however, take shelter in rock clefts, dense vegetation or tree cavities; while some, such as the wombats (*Vombatus*), dig deep galleries similar to those of the placental badgers and marmots. The Duckbilled platypus (*Ornithorhynchus anatinus*) also excavates deep burrows in river banks, making a nest of wet grass in which the eggs are laid.

Some habitations are extremely complex and perhaps none is more fascinating than those of the beavers (*Castor*). Their lodges are constructed on the side of a lake or on a small island. The entrance to the lodge is under water, but at the top there is an air-hole to the inside chamber. The water level is kept steady by dams of wood and debris.

For all mammal species that dig burrows or build nests, these homes represent the heart of their own territory. Even when a mammal does not resort to any permanent shelter of this nature, it occupies its individual territory. This consists of an area which it will defend and from which it will bar other animals of the same species. The only exceptions to this rule are the ocean mammals. A territory is usually marked by means of odorous substances produced by various types of scent glands, which are well developed and particularly common among mammals. The animal generally rubs the secretion from the cutaneous scent glands against prominent natural features within its territory, such as tree trunks, branches and the like. Alternatively it leaves traces of its scent on the ground among rocks or through tangled vegetation. Although every animal has its own territory, some exhibit territorial behaviour patterns throughout the year, while others, including deer, only make such displays during the mating season. The majority of carnivores mark their territory with urine, but some species use their own excrement. Rhinoceroses stake out territory with piles of dung, and the hippopotamus rotates its tiny tail at exceptional speed, spraying excrement all around in almost liquid form.

▼ Herd of hippopotamuses (*Hippopotamus amphibius*)

Territory, however marked out, is recognised as such by strangers of the same species who avoid approaching it. In addition to territory which is respected both by rivals and enemies, there is also a surrounding area, sometimes quite extensive, where the animals hunt and forage for food. This region, or home range, may often overlap those of other individuals or groups to form a neutral zone. Territorial behaviour, apart from playing a vital role during the mating season, is closely associated with the never-ceasing quest for food, and takes on differing patterns according to whether the animals are herbivores or carnivores. The former, should they be small creatures such as fieldmice and other rodents, feed in the immediate vicinity of the burrow, and do not need to stray far afield. The large ungulates, on the other hand, are constantly on the move in search of fresh pastures. The carnivorous predators depend on an ample supply of protein, and satisfy their needs with well spaced out meals. They do not therefore, have to spend the major part of their time feeding, and their territorial behaviour is a good deal more complex, as is their entire range of individual and

▼ Striped skunk (*Mephitis mephitis*)

▲ Female Guinea baboons (*Papio papio*) and young

social behaviour. They indulge in a variety of activities, including play, which are specifically designed to reinforce the social links between the members of the community. Some carnivores set aside part of their prey to be eaten later at intervals, but this pattern of food storage is more common among animals such as rodents, which consume seeds or dry fruits. Many of these animals collect grass and seeds during the summer months to provide sufficient food for winter. The North American pocket gophers (Geomyidae) and Eurasian hamsters (Cricetidae) construct proper storerooms in special chambers of their underground burrows where they accumulate large quantities of food, often up to several kilos in each burrow.

Some forms of animal behaviour are clearly related to self defence, reproduction and the acquisition of food. Mammals, however, often engage in a wide range of other activities. Apart from play, which helps to cement social links, there are, for example, many rituals related to care of the coat. A number of mammals devote themselves to grooming activities designed to keep skin and fur in good condition, and some species are equipped with specialised fingers that can be used effectively as combs. People who keep pets will be familiar with the behaviour of cats who fastidiously lick their fur, and of hamsters and other small rodents who keep themselves clean in a similar fashion. Among ungulates, too, mothers can often be observed licking the skin of their young.

The grooming habits of monkeys are particularly interesting. In their case, these rituals play an important part in maintaining links between different members of the group, and can be observed both in the wild and in captivity, as anyone visiting a zoo can testify. However, not all gregarious animals behave alike with regard to physical contact. Some species appear to avoid it as much as possible, each individual keeping a minimum distance from its neighbour. Others prefer to crowd together, pressing up closely to one another. Seals, walruses and sea elephants will pack tightly together when they come ashore to mate.

Bats which spend the winter in caves exhibit different types of behaviour, according to species. Sometimes individual bats can be seen hanging upside down, wrapped in their wings but a good distance away from one another. Some position themselves closer to their neighbours, though not in actual contact, and others form densely packed clusters. One reason for this close contact may be to help keep the body temperature at a consistently high level. This probably explains why young bats press so tightly against their mothers and siblings at a period of life when the thermoregulating mechanisms are not yet functioning properly. In other cases such behaviour would appear to fulfil a psychological need.

Reproduction

In mammals, as in all vertebrates, reproduction is of the sexual type, by means of specialised single cells known as gametes, produced by the female (ovum) and the male (spermatozoon). These unite in the process of fertilisation to produce a new individual. Each gamete contains the hereditary material of the individual produced. As these particular cells mature, there is a process known as meiosis whereby the chromosomatic set of each ovum and spermatozoon is reduced to a haploid (single) state, in which there is only one chromosome for each corresponding pair of chromosomes present in either parent. When the male and female gametes come together, the complete set of chromosomes is restored to the diploid (double) condition characteristic of that particular species. In this way, each new individual that is formed possesses one set of chromosomes from the father and one set

67

from the mother, so that the genetic code present in the chromosomes can combine in an almost unlimited number of ways. This means that there can never be two exactly identical individuals unless they originate as the result of the division of one diploid cell (the fertilised egg or zygote), as in the case of single-ovum or identical twins. Twins developing in this manner from the same and not from different ova are a comparatively rare occurrence, but it is, of course, known among humans and is a normal phenomenon in the case of some species of armadillo, when four, eight or even twelve identical twins, all derived from the same egg, may be produced.

The three essential stages of reproduction are fertilisation, in which the male and female gametes unite; the development of the embryo, entailing the division of the fertilised egg and the multiplication of cells to produce a new individual, together with various methods of protecting and feeding that embryo; and, finally, birth and post-natal development.

The egg cells are derived from the outermost layer of the female gonads (ovaries) and are situated inside a small bladder-like cavity, the Graafian follicle. When the egg matures this follicle fills with fluid, the wall extending and projecting above the surface of

▼ Female red kangaroo (*Macropus rufus*) with young in marsupium

the ovary. During ovulation the follicle bursts and the egg enters a kind of ciliated tunnel (the uterine or Fallopian tube) which continues in the form of a narrow canal (oviduct) leading to the uterus. It is generally in the oviduct that the egg encounters the spermatozoa, introduced as a result of copulation, which swim up the vaginal canal, the uterus and the oviducts by lashing their tails spirally. The spermatozoa are produced by the male gonads (testes), which are made up of a large number of tubules. The mobile sperm cells detach themselves from the walls of these tubules and enter a spirally wrapped tube called the epididymus where they are stored. This leads to the exterior by way of the vas deferens. There are two testicles or testes, situated either inside the abdomen or in an external sac which may be permanent or temporary. There are two ducts which open out on either side into the urethra, which functions as an ejaculatory duct.

In addition the male possesses other accessory glands that produce thin fluids in which the spermatozoa can move about and so enter the female vagina as a result of copulation. In some mammals these fluids have the capacity of clotting after entering the female, so as to form a kind of plug that serves to hold back the sperms, thus making fertilisation easier. This happens, for

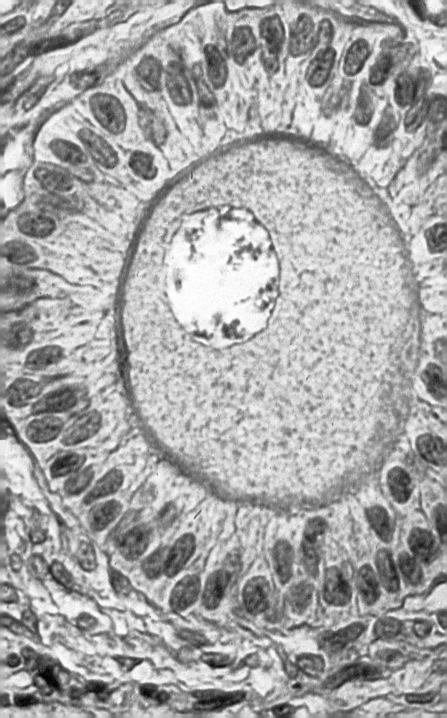

example, among many bats, where fertilisation takes place several months after copulation.

In the course of fertilisation the spermatozoon penetrates the ovum, and the nucleus (contained in the head of the sperm itself) unites with the egg nucleus, thus forming the diploid zygote which is the first cell of the future individual. Chemical modifications of the egg surface at the moment of fertilisation prevent any further penetration by other spermatozoa. At this stage the zygote begins to divide into an ever-increasing number of cells which constitute the various tissues and organs, according to the development of the particular genetic code.

In nearly all mammals, therefore, the ovum is fertilised inside the mother's body, and almost always in the uterus or womb. Only in the case of the monotremes is the fertilised egg laid in the manner of reptiles and birds, developing outside the mother's body. There are, therefore, certain differences between monotremes and other mammals. The eggs are larger and the

▼ Section of epididymis, the duct linking the testicle to the vas deferens

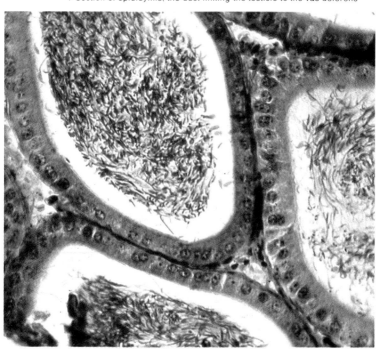

◄ Graafian follicle inside a mammalian ovary

oviducts are two simple tubes, one on either side, which at their lower end terminate in a urinogenital apparatus opening into a cloaca. Apart from this, only the ovary and oviduct on the left side are functional, as in birds, and the egg is fertilised in the part of the oviduct that virtually corresponds to the uterus. After it is covered by a shell the egg is laid. It is either incubated in a nest (as with the Duckbilled platypus) or in a kind of pouch (echidnas). The marsupials have a double vagina joined at the top, at the link of the two uteri. At the bottom they merge into a single vagina which opens out, together with the excretory and digestive systems, into a cloaca. As a result, the tip of the male penis is split into two parts. The testicles are enclosed in an external sac positioned in front of the penis instead of behind, as in all other mammals.

Another difference is the way in which the young, which are virtually immature embryos, are born. Instead of finding their way into the world via the lateral vaginas, they are born through a temporary orifice that links the mouth of the uteri (at the upper juncture of the vaginas) with the opening at the lower end. In some kangaroos this passage may become permanent after the first birth. In marsupials, as in placental mammals, the development of the embryo and its appendages takes place inside the uterus, an organ here symmetrically present on both sides of the body. In the placental mammals this organ is capable of considerable enlargement, and in some cases (certain bats and primates) the two uteri are fused to form a single median uterus, while in other orders only the extremities are fused. The lower end of the placental uterus leads to a vagina that opens to the outside. It is through this orifice that the seminal fluid flows during copulation, and it is through this channel that the young are born.

There are obviously a number of structural differences between the reproductive organs of the various groups and even species of placental mammals. Sometimes the urethra (the last section of the excretory apparatus) has a common outlet with the vagina, but sometimes there are two distinct openings. Occasionally the vaginal aperture forms only after sexual maturity or, as in moles, is present only during the reproductive period, closing after birth. In spite of these differences, once fertilisation is accomplished the zygote attaches itself to the uterine wall which in the meantime has thickened markedly as a result of the action of progesterone. This hormone is secreted by the follicle cells of the ovary (transformed into a ductless gland called the corpus luteum, following ovulation). The external cells of the ovum after cleavage (blasto-cysts) form a membrane, the chorion, which envelops the embryo and adheres to the uterine wall. In placental mammals the

▲ Female gorilla (*Gorilla gorilla*) in advanced state of pregnancy

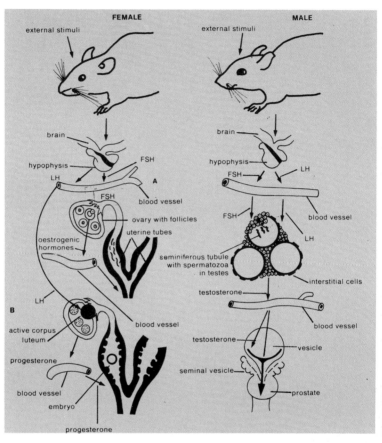

▲ Hormonal control in the reproductive cycle. Female: (A) FSH, follicle-stimulating hormone causes ova in follicles to ripen and be discharged; The follicle cells produce oestrogenic hormones which are converted to corpora lutea by LH, Luteinising hormone (B). Male: FSH, hormone stimulating spermato genesis; LH, interstitial cell-stimulating hormone (See p. 76)

link between the uterine lining and the chorion of the egg becomes more complex and forms the placenta. This is an organ developed so that the embryo can feed and breathe. The structure is rich in vascular tissue, enabling the blood vessels of the mother and the embryo to come together in fairly close contact (depending upon the type of placenta) without any mingling of blood supplies. The structure of the placenta varies according to order and even according to genera, but as a rule the basic pattern is the same,

▲ Courtship of entellus monkeys (*Presbytis entellus*)

resulting from the union between the chorion and the uterus.
Other constituents may include the blood vessels of the vitelline
sac or yolk (the most primitive or omphaloid type of placenta,
also found in marsupials) or the blood vessels of the allantois (the
more evolved allantoid placenta, typical of eutherians and some
marsupials). The placenta remains attached to the wall of the
womb throughout gestation and is linked to the navel of the
embryo by the umbilical cord, formed from the allantois and

75

vitelline sac. At birth the cord is cut, and in the majority of eutherians the placenta, or afterbirth is expelled following birth (parturition). In some species the embryonic tissues and the placenta itself are eaten by the mother, in others the placenta is abandoned. In marsupials and certain insectivores it is resorbed by the uterus itself.

The length of gestation is extremely variable. Among marsupials it is short, ranging from a couple of weeks in the opossum family to between four and six weeks in kangaroos. The newly born marsupial is very backward in its development. Because of its small size it cannot live far from its mother. Despite its immaturity it manages to travel from the vaginal opening as far as the marsupium (pouch) by means of swimming movements and by gripping with its well-developed forelegs. Depending on the species, this pouch may be in various positions, sometimes large (red kangaroo), sometimes small (koala), occasionally absent (banded anteater). Either inside the pouch or in the fur there are nipples. The young marsupial is firmly attached to the nipple, which swells in its mouth, for the whole period of lactation.

Mating behaviour

The overall behaviour of animals during the mating season and the activities of their internal and external organs are under the complex control of chemical substances called hormones. These stimulate or inhibit particular reactions and together with the endocrine glands constitute a delicately balanced system. The table on p. 74 shows the essential hormonal influences on the reproductive cycle, both in the male and female. The gonads only produce sexual hormones if they are stimulated by gonadotropic hormones, which are developed in the forward part of the hypophysis or pituitary gland of the brain. These hormones develop the follicles of the ovary and stimulate the secretion of oestrogen. The luteinising hormone encourages the formation of the corpora lutea after the extrusion of the follicles and stimulates the production of progesterone. Corresponding to these female hormones is a male hormone secreted in the tubules of the testes which stimulates sperm production, and another, similar to the luteinising hormone, which acts on the interstitial cells. The pituitary gland also produces oxytocin, causing the muscles of the uterus to contract during labour, as well as prolactin to stimulate the flow of milk in the mammary glands after birth. The pituitary is stimulated into producing the gonadotropic hormones by a variety of outside factors. These include the amount of light and the duration of the day, which would explain why sexual activity tends to be seasonal. Other contributory factors are changes in the environment, or modifications in behaviour patterns, such as courtship displays.

Smell is one sense that has a major influence on sexual behaviour. Many scents have a stimulating effect and have the same function as pheromones, the social hormones known to exist in other groups of animals, particularly insect communities. This can be observed especially in wild carnivores and also among domestic cats and dogs, all of whom are attracted by the characteristic odour exuded by the secretions of their partner of the opposite sex. These scents given off by the female trigger a response in the male leading to the behaviour necessary for copulation. This pattern of behaviour may take various forms. In deer, for example, the belling of the male and the combats between rivals both provide a stimulus to the female. The majority of ungulates put on courtship displays that involve rearing up and other movements that serve to accentuate secondary sexual features such as manes, horns and coat colour. The displays of Uganda kobs (*Kobus kob thomasi*) are very striking. In the course of these the males display the white patches on throat and chin, and the black stripes on their legs.

▼ Red deer (*Cervuselaphus*): a stag with his harem

In virtually all the ungulates, the male, prior to mounting, touches the female several times with his stiffly held forelegs. This happens among the Uganda kobs and also among the oryxes. Oryxes also preface mating with brief combats in which the male asserts his dominance over the female.

In the case of gregarious species and in those that do not observe a determined mating season, there is no such courtship display or rivalry. For example, female elephants in heat couple with the dominant male and with other males as well. Primates

▼ A pride of lions (*Panthera leo*) in the savanna

provide a field of rewarding research into reproductive behaviour if only because the different species reveal a wide range of social conventions. The gibbons (*Hylobates*) form stable pairs whereas the sacred baboons (*Papio hamadryas*) have a harem organisation. Intermediate situations are to be found in other primate societies. In almost all monkey species it is the female who arouses the interest of the male for purposes of copulation. Many females exhibit bright patches of colour, particularly in the genital regions, which act as visual stimuli.

▲ Female llama (*Lama glama*) suckling young

Among the Canidae some species, such as the fox (*Vulpes vulpes*) and the wolf (*Canis lupus*) form pairs that last at least for the whole reproductive season. In these cases the male helps to rear the young and brings food to the female while she is suckling. Among the Felidae, male and female generally meet only during the period of oestrus and the courtship display is mainly vocal. In lion communities, however, there is no rigid social structure and the animals form polygamous family groups.

Most of the smaller rodents and the insectivores do not appear to go in for courtship display. In their case the period of oestrus is very short, and mating occurs after brief pursuits by the male. The female will normally couple with several males in succession. However, the hedgehogs (*Erinaceus europaeus*) perform a lengthy display in which both partners engage in combat, striking each other with the spines of the forehead until the female is receptive to the sex act.

Other important patterns of behaviour relate to the care of the young, both before birth, during labour and after the young are

▲ Female wolf (*Canis lupus*) with cub

born. One of the more primitive mammals, the Duckbilled platypus, digs a suitable burrow, about 12 m long. In the burrow it builds a large grassy nest for the incubation of the eggs. This is where the female suckles the young for the first weeks of their life.

The problem of protecting the babies is easily solved in the case of all other monotremes and marsupials, usually by virtue of the marsupium. Where a pouch is not present, the young have to cling firmly to the nipples situated among the hairs of the stomach. The number of nipples often determines how many youngsters will survive if the litter is a large one. Female marsupials do not appear to provide any special treatment for their offspring during the actual birth. The young have to find their own way to the pouch or to the nipples, and only on rare occasions will the mother trace out a path with saliva. Should the young animal fall, it is simply abandoned to its fate. The koala (*Phascolarctos cinereus*) behaves somewhat differently, however. Once the young have left the marsupium, they are transported on the mother's back. Before they begin eating leaves on their own, they are fed on pieces of her excrement, which consist of the partially digested pulp of eucalyptus leaves.

Among placental mammals the mother often prepares special nests or burrows, where she awaits the birth of her offspring. This almost always occurs in rodents, whose nests accommodate mothers and young during the period of suckling. The nest of the rabbit is placed at the bottom of a special burrow (the stop) and is lined with dry leaves and hair. When she has to leave the young on their own, the mother closes up the entrance with earth. Insectivores behave in a similar manner and carnivores construct a type of nest in an excavated burrow or in a cavity. This can be either natural or artificial, depending on what is available. With the exceptions of the fox and the wolf, it is the female carnivore who prepares the nest and looks after the young on her own after they are born. Ungulates, such as antelopes, do not usually make any special arrangements for the new-born baby who is capable of following the mother or the herd almost immediately. Just prior to giving birth female gazelles and deer generally wander off alone to look for a safe shelter where the young can be left for the first few days on a litter of grass.

The period during which the baby takes milk from the mother (even in gregarious species it is only the mother who suckles her own offspring) varies considerably from one species to another. When the young animal is eventually weaned it is generally capable of obtaining food for itself. In the case of carnivores, however, the mother initially has to feed her cubs with meat from the prey that she has caught. At that stage the young lack the strength and experience to hunt for themselves and have to learn how to hunt by imitating their parents. Play activity is of the utmost importance to young carnivores as it teaches them the necessary techniques of hunting and fending for themselves. Lion cubs learn to hunt not only with their mother but also with other members of the pride. These are usually young females from previous litters or other lionesses without cubs. Brown bear cubs, generally born in pairs in the lair during the winter, also learn to hunt with their mother. They make up a small family group, while the father leads a solitary life. Such groups include cubs born the previous year, provided they are less than eighteen months old.

American black bear (*Ursus* ▶
americanus) with cub

▼ Lions (*Panthera leo*) feasting on zebra

▲ Walruses (*Odobenus rosmarus*)

Populations

In the preceding sections, mention has been made of territorial and gregarious animals, but with an emphasis on social organisation. Reference has also been made to groups of individuals belonging to the same species, which are known as populations. Among the more solitary species, in which individuals occupy large territories and meet only during the mating season, one can regard a population as an assembly of animals of a single species, living in a given area. The most valid definition of a population, now accepted by most biologists, is that of Ernst Mayr, who describes it, in a strictly local sense, as 'a community of individuals capable of reproducing by cross-breeding within a given locality'. It is impossible here to discuss the genetic variations of populations but the following section briefly examines the sizes of populations and the fluctuations in their numbers, which are collectively known as their dynamics.

In order to make a proper study of any animal population, the first requirement is to effect a census to find out how many individuals exist in a given area. In the case of large mammals, the only method is to observe and count them as they pass through a particular area. Fairly successful results can be achieved by aerial photography, especially with ungulates of the treeless savannas, or groups of seals on beaches during the mating season. Where small mammals are concerned, various methods have been used which give reasonable approximations but they are all difficult to put into practice. Theoretically, the simplest procedure would be to catch all the individuals of a population and to count them, but obviously this is virtually impossible.

In practice, the method generally employed – not just for mammals but for almost any animal population – is that of marking and recapture, and the simplest estimate of a population is derived from the Lincoln Index. The formula is:

$$N = \frac{X_1 \times X_2}{Y} .$$

Where N is the total population in an area, X_1 is the number caught, marked and released at the first sampling; X_2 is the total number caught, marked and released at the second sampling; and Y is the number of the X_2 animals that had been marked at the first sampling. There are problems involved even with this method, and it is often necessary to repeat the operation several times. Furthermore, the animals that have been captured more than once learn from experience, often becoming bolder and managing to remove food from the traps laid for them. Appropriate calculations based on this method can also provide information about population changes and replacements, length of life and so forth. Despite the difficulties presented by these methods, as well as by other 'selected sample' systems, zoologists are able to assemble much valuable information concerning the dynamics of animal populations.

The main reason why fluctuations in numbers occur is that the amount of food varies. Sometimes swift and irregular changes in numbers can also be caused by periods of severe drought or by epidemic diseases. Populations of certain small mammal species (particularly rodents) may be subject to regular cycles of increasing and decreasing numbers. Some types of fieldmice, for example, show an enormous increase in numbers every three or

Overleaf: herd of common zebras (*Equus sp.*) ▶

▲ Alaska fur seals (*Callorhinus ursinus*)

four years, as a result of the speed with which they reproduce and a low mortality rate. When population density reaches its peak, there is a sharp and sudden fall in numbers and eventually population density reaches its minimum. The reasons for such fluctuations among fieldmice are complex, and are now known not to be associated with disease or food shortages. Evidently they are related to the high density level itself which leads to an increase in aggressiveness and stress factors, thus causing an endocrine imbalance as well as a decrease in fertility and a drop in the birth rate. With the population at its lowest level as a result of

Field vole (*Microtus arvalis*) ▶

▲ Norwegian lemmings (*Lemmus lemmus*)

deaths by stress, the few survivors begin breeding again and the population increases.

One of the best known examples of this phenomenon is that of the Norwegian lemmings (*Lemmus lemmus*) who undertake massive emigrations about once every four years when the food resources of their mountain habitats have dwindled almost to nothing. This is a signal for them to make for the plains, finding their way across roads, paths, lakes and rivers until they die of starvation and exhaustion. The comparatively few individuals who do survive the catastrophe return to their place of origin and help build up the population to its maximum level. This in turn leads to new crises and fresh disasters. There are also similar population fluctuations among the predators of such species. The lynxes and foxes of North America undergo variations in number which are correlated with those of the snow hare (*Lepus timidus*) which they hunt and which reach a maximum density level every ten years. Spectacular fluctuations in numbers also occur in species of small animals with a rapid reproductive cycle. These are, however, merely extreme examples of phenomena to be found

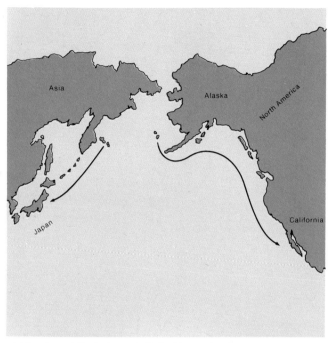

▲ Migrations of Alaska fur seals (*Callorhinus ursinus*)

in virtually all mammals, though in other species the population changes are less evident. All depend upon the availability of food, the number of predators in the area and, sometimes, on the actions of man who, by eliminating predators, may create an imbalance between populations and habitat.

Migrations

The term 'migration' is commonly used to describe movements of populations that occur more or less regularly. These movements generally entail an outward and return journey at fixed periods, and are either seasonal or annual. Although many herbivores of the African savanna move off in search of new pastures enriched by the seasonal rains, there are comparatively few species of land mammals that make regular migration journeys. The reason is simply that such journeys are too difficult on foot. Migration is, therefore, a far less common phenomenon for mammals than for birds, although mammals that can swim or fly are capable of

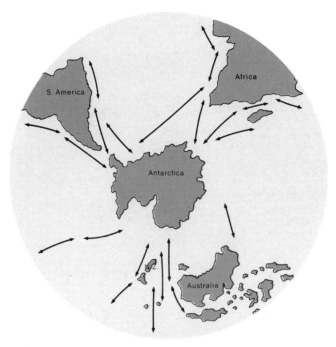

▲ Migrations of humpback whales (*Megaptera novaeangliae*)
from feeding zones in Antarctica to breeding grounds

undertaking longer migrations, as shown in the above charts. The
Alaska fur seal (*Callorhinus ursinus*) breeds in the Islands of the
Bering Sea, where the adult males stake out territory and
assemble harems of females. In the autumn, when the suckling
period is over, the groups of seals return south; the adult males
moving towards Japanese and Canadian waters, the females and
young return to areas around Lower California, more than 4,800
km away. Similarly whales, such as the humpback (*Megaptera
novaeangliae*) regularly travel enormous distances. The
humpback whales, as well as some rorquals, arrive off the
Antarctic coasts in summer when the sea is swarming with the
plankton required to maintain their reserves of fat. In winter they
travel back to tropical and subtropical seas, where the young are
born. It is not yet entirely clear as to how such mammals navigate
during these migrations.

As far as flying mammals are concerned, certain species of
American bats make regular migrations from north to south and

vice-versa, like many birds. Other species also undertake long journeys, but not regularly, usually spending the winter months hibernating. As already mentioned the land mammals that do regularly migrate include antelopes and gnus that trek across the African savannas, and the saigas (*Saiga tatarica*) of the Asiatic steppes. Two other species that make long migration journeys are the caribou and bison of the North American prairies. The caribou migrate regularly each year, displaying an accurate sense of direction. They move in huge herds, leaving the Canadian tundra in the autumn and making for the taiga. The following spring they retrace their steps, arriving back in the tundra just when new plant life begins to appear. Here they can feed on the fresh grass and on the leaves of dwarf willows. Here they also give birth to their young. The mass migrations of the American bison (*Bison bison*) or buffalo across the North American prairie were once a spectacular sight. In the 19th century this bison population exceeded 60 million head. These magnificent beasts made regular journeys between the Rocky Mountains and the eastern forests. They travelled not only from north to south, but also from east to west, covering hundreds of miles on a more or less circular course.

▼ Reindeer (*Rangifer tarandus*)

On such migrations, the buffalo were tracked by wolves and bears, as well as Indians who depended on the animals for their very existence. Because of the enormous size of the buffalo herds and the fairly primitive hunting weapons then available the Plains Indians only killed a small proportion of the total number of buffalo and the situation was kept in balance. Between 1860 and 1880, however, the white man deliberately exterminated the buffalo herds, not only for the purpose of securing skins and meat, but also to threaten the survival of the Indians. Saved from extinction at the last possible moment, the buffalo population today numbers only a few thousand head. These animals lead what is virtually a sedentary existence in several reserves.

Various other mammals undertake less dramatic migrations, travelling down from the high mountain regions where they live in the summer to the valleys in winter. Such animals include the red deer, the roe deer, the chamois, the alpine hare, the fox and the marten. Domestic mice, which live out in cornfields in the summer and move back into barns and lofts in winter, can also sometimes be said to migrate, even though the distances they cover may be measured in metres rather than kilometres.

▼ American bison (*Bison bison*)

Dormancy

Migrations enable many mammal populations to resist the
unfavourable conditions of their environment, by moving to more
suitable regions at fixed times of year. Some animals living in
regions subjected to climatic extremes survive in a different
manner, by entering into a state known as dormancy. If this
happens in winter this state is called hibernation, while summer
dormancy is known as aestivation. During dormancy the body
metabolism drops to its minimum level and all the bodily
functions are slowed down. Few mammals go through a phase of
dormancy in its proper sense, and of those that do, the majority
are bats living in temperate climates. Only a few North American
species fly south in the winter, the others sheltering in tree cavities
or in caves, often in enormous groups, for the entire season.

In practise, dormancy is a state of torpor in which the
temperature of the body, controlled by the nerve centres, is
significantly lowered. A mammal in this condition never loses its
capacity of thermoregulation, however, and if the temperature of
its environment becomes excessively low, it will react by waking
up. Before becoming dormant the species concerned stocks up
with body fats. These consist of deposits of white fat (an energy
reserve which serves to maintain the reduced rate of metabolism
during the period of torpor) and brown fat (which produces the
calories necessary for reawakening). It is not only external
circumstances, such as shortage of food and falling temperatures,
that induce dormancy. Internal conditions controlled by
hormones are also influential. For example, the quantity of insulin
regulates the amount of sugar in the bloodstream and probably
affects the thermoregulating centre of the pituitary gland. During
dormancy there is also a notable increase in the quantity of
adrenalin produced by the surface layer of the suprarenal glands
and a reduction in the amount of noradrenalin. It is not known
exactly what part is played by the different hormones in
controlling temperature and stimulating dormancy, but this state
is evidently triggered off by the connection and interaction of
adrenalin, noradrenalin and the thyroid hormone. The principal
agency bringing about torpor has been proved to be insulin. By
injecting the substance experimentally into hedgehogs a state of
artificial lethargy is caused. Another physiological change linked
with this condition and with the lowering of body temperature is
an extreme reduction in breathing activities. In addition the heart
beat rate is reduced, resulting in a slowing down of the blood-
clotting process which avoids the danger of thromboses.

Reawakening may be provoked by a gradual rise in outside
temperature, by some form of mechanical stimulus or, as already

▲ Common dormice (*Muscardinus avellanarius*)

◀ Edible or fat dormouse
(*Glis glis*) hibernating

95

▲ Hedgehog (*Erinaceus europaeus*)

mentioned, by a lowering of temperature in the immediate
environment. As a consequence of such stimuli the rate of
heartbeat increases, the body temperature rises and the metabolic
rhythm is restored to normal. The dilation of blood capillaries and
movements of certain muscles (causing shivers) are important
factors in bringing about this rise in body temperature.

Instead of hibernating, some rodents and insectivores do
precisely the opposite. Those which live in hot regions where
summer temperatures are intolerable, resist heat and drought by
aestivating. There are also a number of small mammals that
undergo phases of daily torpor, manifested in a lowering of the
rate of metabolism during the cold hours before dawn or when
there is a food shortage.

Apart from bats, there are relatively few European species of
mammal that hibernate in the true sense. Among insectivores,
only the hedgehog goes through a period of dormancy which lasts
no longer than three months. Even this, however, tends to be
irregular, punctuated by periods of wakefulness. The best known

examples of hibernation in rodents are to be found in the Gliridae. Both the common dormouse (*Eliomys quercinus*) and the fat or edible dormouse (*Glis glis*), build summer nests in trees, and also make underground nests in winter. Before becoming dormant they grow extremely fat and then go to sleep by rolling up into a

▲ Alpine marmot (*Marmota marmota*)

ball, face resting on the stomach and feet folded on either side. While asleep their body temperature is very low and the muscles are stiff. They look exactly like balls of fur and can easily be rolled about without changing position or waking up. This period of dormancy may last for more than five months. In Europe squirrels do not actually hibernate and are active all year round, but the alpine marmot (*Marmota marmota*) is dormant for up to eight months in winter. These periods of dormancy are spent inside large, elaborate burrows.

Other mammals which hibernate in cold or temperate climates include certain species of marmots in the Holarctic region, some ground squirrels (*Citellus*) and the so-called prairie dogs (*Cynomys*). Some animals alternate periods of dormancy with intervals of activity, during which they eat food accumulated in their burrows. These are constructed deep underground to withstand frost and extremely low temperatures.

All European bats spend the winter in sheltered places and many live in caves. No matter how severe the cold outside, the

▲ American black bears (*Ursus americanus*)

temperature inside the cave is always above 0°C. The bats do not sleep throughout the winter months, for many species wake up at irregular intervals and may even move about, either within their cave or from one cave to another.

It is often assumed that the brown bear (*Ursus arctos*) is a hibernating creature. It is true that in the more northerly areas of distribution it will shelter in a lair for several months. However, it never hibernates in the full sense, as its body temperature remains normal even though the heart-beat may frequently slow down. During this period the female gives birth to her cubs and suckles them in the cave for a few weeks. The female polar bear (*Thalarctos maritimus*) behaves in a similar manner, excavating a lair in the snow, where she gives birth and rears her young.

◄ Horseshoe bat (*Rhinolophus ferrum-equinum*)

Habitats

The subdivision of the earth into zoogeographical regions, as accepted by modern scientists, is based on the scheme introduced more than a century ago by Alfred Russel Wallace, the distinguished contemporary of Darwin who shares the credit for the theory of evolution. Wallace distinguished six major zones on the basis of their floral and fauna composition – the Palearctic, the Nearctic, the Neotropical, the Ethiopian, the Oriental and the Australian. Because of its isolation, the island of Madagascar (Malagasy) is often treated as a separate region. It is, however, sometimes linked with the Ethiopian and Oriental regions to form the so-called Palaeotropical region. Polynesia is also regarded as a separate region. There are many common features to be found in the Palearctic region (Europe, Asia north of India and Africa north of the Sahara) and the Nearctic region (Greenland and North America down to central Mexico) and as a result they are often treated as a single Holarctic region.

Many mammal species have a Holarctic distribution. These include the polar bear, the wolverine or glutton (*Gulo gulo*), the wolf, the Arctic fox (*Alopex lagopus*), the reindeer and the elk. Alternatively, they are represented in Eurasia and North America by different races, such as the brown bear, the red deer, and the fox, or by slightly variant forms, like beavers, bison and minks. The Holarctic region has many interesting species of insectivores, bats, lagomorphs and rodents, carnivores (including many mustelids) and artiodactyls. These last include many Cervidae

and goat-like Bovidae, with certain species adapted exclusively to life in high mountain regions. One such family, the Antilocapridae, is found only in North America where there are no primates and perissodactyls. The opossum is the sole representative of the marsupials in the Nearctic region, and is a comparatively recent arrival from South America.

The Ethiopian region, comprising Africa south of the Sahara, contains a large number of Bovidae and many endemic species of carnivores, rodents and primates. There are no marsupials or edentates, but the region does have representatives of the Hyracoidea (hyraxes) and Proboscidea (elephants). Native to the Ethiopian region is the order Tubulidentata (aardvarks) and the families Hippopotamidae and Giraffidae. The fauna of Madagascar is very specialised, with a number of indigenous animals including many Enosimii (Lemuridae, Indridae and Daubentoniidae – lemurs, indris and aye-ayes), some carnivores (Viverridae), insectivores (Tenrecidae) and rodents.

The Oriental region, comprising southern Asia and the archipelagos of Sunda, Borneo and the Philippines, is separated from the Australian region (consisting of Australia, Tasmania, New Guinea and the islands around Celebes), by the so-called 'Wallace line'. This imaginary line, very distinct in the case of plant life, but less so for animal life, passes between the islands of Bali and Lombok. In the Oriental region there is only one indigenous order, the Dermoptera (flying lemurs). However,

▲ The world's zoogeographical regions

	Nearctic region		Ethiopian region
	Oriental region		Neotropical region
	Palearctic region		Australian region

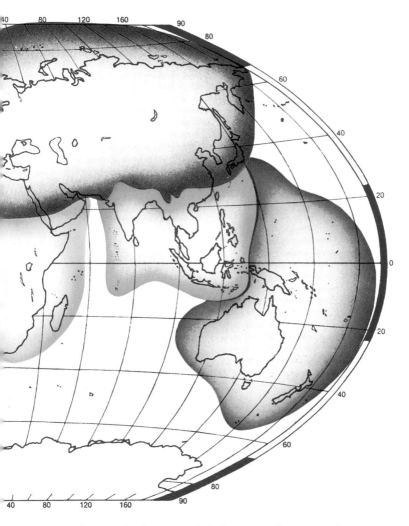

there are also hyraxes and elephants as well as many primates, carnivores, artiodactyls and perissodactyls.

The Neotropical region, made up of South America, central America and the Antilles, is characterised by several endemic groups including the entire order Edentata (anteaters, sloths and armadillos), the marsupials of the families Didelphidae (opossums) and Caenolestidae (pouched shrews or opossum rats), the platyrrhine or New World monkeys and a number of rodent families. There are only a few insectivores.

Of all the zoogeographical zones, the Australian region is probably the most individual, for its animal population consists of the indigenous order Monotremata and of the majority of the world's marsupials. Because of its isolation since very ancient times (the Cretaceous), the placental mammals were not able to colonise it on a large scale. The present day mammal population of this region results from adaptive radiation of the original species. These species are specialised to cope with a variety of living conditions. In scientific language, they can be said to occupy a wide range of ecological niches. Only a few eutherians have been able subsequently to settle in the Australian region notably seals and whales as well as some bats and a few rodents of the family Muridae. At one time New Zealand did not have any land mammals, except for two bat species. In the 11th century however the Polynesians introduced the dog and the rat and after 1779, Europeans brought in a range of animals including elks, chamois, flying squirrels, hedgehogs and rabbits.

Having surveyed the zoogeographical regions, the following section deals with the distribution and life styles of the mammals occupying different habitats in these regions. Because of its special nature it is appropriate to begin with the Australian region.

The Australian region: the mammal-reptiles (Prototheria)
The monotremes are the only order of the subclass Prototheria and they are without question the strangest of all mammals. This curious order consists of two distinct families – the Tacnyglossidae (echidnas) and the Ornythorhynchidas (Duckbilled platypus). Some of the chief mammal characteristics of monotremes have already been mentioned (hair, mammary glands – though specialised and lacking nipples, and a brain similar to that of marsupials). Other characteristics are typical of reptiles (shoulder girdle bone, a reproductive apparatus with a cloaca and, above all, oviparity). Clearly this is a very archaic group which has survived only because Australia, Tasmania and New Guinea were isolated from other continents. Therefore, it is believed that the monotremes do not represent a stage of mammalian evolution, as in the case with marsupials and eutherians, but a parallel line of evolution which broke away very early from the original mammalian stock. No ancient fossils of monotremes have been found and the present-day species exhibit very specialised adaptations relating to their modern habitats. The muzzle has lengthened into a kind of beak, while the adults lack teeth (only the maxillary teeth are present in newborn platypuses) and have instead horny plates on jaws and tongue. The feet are very specialised for digging.

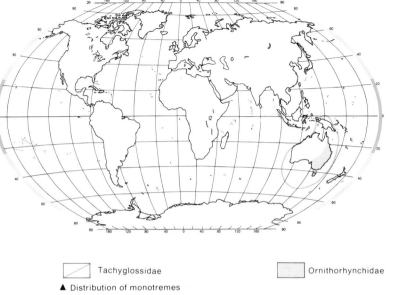

| | Tachyglossidae | | Ornithorhynchidae |

▲ Distribution of monotremes

The echidnas are medium sized animals with a body length of up to 80 cm. They possess a number of fairly long, pointed spines, which mingle with the hairs of the back. They have a very short tail and their feet are furnished with large claws. At the tip of the long, beaked muzzle is a small mouth with a slender tongue that darts rapidly in and out. This long tongue is covered with sticky . saliva that enables insects to be caught. The echidnas feed mainly on ants which they crack open with the horny plates of the palate and the base of the tongue. The echidnas of the Australian region are the ecological equivalents of the South American anteaters and the African aardvark. As with their counterparts, echidnas use their powerful feet to break open anthills and termites' nests and to lift trunks and stones. The slender muzzle, which has nostrils at the tip, is ideal for searching out insects sheltering in these spots. They also use their strong claws to burrow in the ground at incredible speed, either to defend themselves against sudden dangers or to provide protection from intense heat. They do not excavate burrows in which to live, but prefer to make use of existing holes either among rocks or tree roots. They can also defend themselves by rolling up like hedgehogs. Echidnas have a long life (50 years or more), and the female lays one egg at a time. The egg is incubated in a kind of pouch where the baby suckles

▲ Australian echidna (*Tachyglossus aculeatus*)

for a lengthy period. The two known species of the genus
Tachyglossus are found in various parts of Australia, Tasmania
and New Guinea. These have a straight muzzle and long spines.

The only member of the other family is the Duckbilled platypus
(*Ornithorhynchus anatinus*). This strange animal lives in streams
and lakes either on the plains or up in the mountains. Its body,
about 30–45 cm long, is covered with thick fur. The tail,
10–15 cm in length, is broad and flattened. The flat muzzle is
shaped like a duck's bill and is covered with soft, delicate skin. It
makes an ideal and highly sensitive organ used for delving
underwater in search of food. Its diet consists mainly of worms,
molluscs, crustaceans and water insects. The webbed feet are
well-suited for swimming and are also useful digging implements.
The males possess powerful horny spurs on their hind feet,
traversed by a channel and linked with a special poison gland.

The Duckbilled platypus excavates long tunnels in the banks
of rivers and lakes. The female uses these galleries as a nest after
lining them with wet grass. Each female lays one to three eggs.

The Australian region: the marsupials (*Metatheria*)

The true mammals (subclass Theria) are today represented by
two infraclasses, the Metatheria, with the single order of
marsupials, and the Eutheria, or placental mammals, with some
sixteen orders. All these mammals are viviparous, and have

▲ Duckbilled platypus (*Ornithorhynchus anatinus*)

distinct openings for the urinogenital system and alimentary canal. They also have mammary glands.

The marsupials (excluding the monotremes) are the most typical and individual mammals of the Australian region. This vast group of animals contains a broad variety of different species which display a similarity to certain placental mammals. The marsupials are not exclusive to Australasia but are also inhabitants of the Neotropical and Nearctic regions, where they are represented by two indigenous families. Despite the numerous and divergent specialised features of present-day marsupials, these animals clearly represent a primitive stage of mammal evolution, in keeping with a pattern that had already emerged in the Mesozoic period. Thus, this order is no mere survival of a side branch, as in the case with the monotremes, but may be a genuinely archaic phase of evolutionary development.

The external appearance of marsupials does not differ much from that of the eutherians, but their reproductive biology is dissimilar in many ways. The young are born in a comparatively undeveloped state, quite unable to fend for themselves, and remain attached to the mother's nipples, usually inside a suitable abdominal pouch or marsupium. They do have some primitive characteristics in common with the monotremes, namely a certain variability in body temperature and a brain that lacks corpora callosa – the structures that link the two hemispheres and which

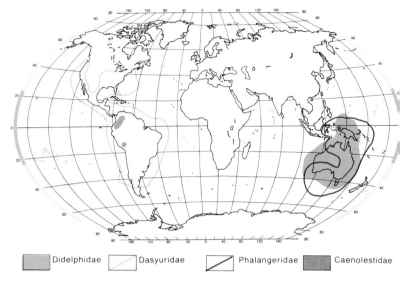

| Didelphidae | Dasyuridae | Phalangeridae | Caenolestidae |

▲ Distribution of various marsupial families

are typical of more evolved mammals. In the skeleton the pelvis supports the epipubic bones, once known as 'marsupial bones', the palate is generally incomplete and the dentition is virtually of the monophyodont type; furthermore, there is only one molar milk tooth.

Of the two modern families of marsupials, six are to be found in Australasia and two in America. One of the latter, the Didelphidae, is probably the closest in resemblance to the primitive marsupials which were once widely spread through Europe and North America. It contains species of varying sizes from that of a mouse to that of a large cat. The animals are for the most part insectivores and carnivores, many of them living in the forests of South America. The American or common opossum (*Didelphis marsupialis*) is one of the best known and least specialised, and is a notable example of evolutionary success. This is certainly the marsupial with the widest area of distribution for it ranges from Patagonia to southern Canada. Other didelphids also have a wide range of distribution but many are strictly localised. Almost all of them, however, are nocturnal animals and expert climbers. The water opossums of the genera *Lutreolina* and *Chironectes* are also, as their name suggests, hardy swimmers. The animals of the latter species are completely adapted to aquatic life, with webbed feet and a marsupium that can be closed by a sphincter muscle.

The other American family, Caenolestidae, comprises a few species in the Andes. All of these small animals live in forests, are nocturnal by habit and feed on insects. They do not have a pouch and have been known in fossil form since very ancient times.

The marsupials of the Australian region are much more specialised for life in a variety of habitats and show a greater range of life styles. In addition to insectivores and omnivores resembling their counterparts in America, there are more highly evolved herbivores, carnivores, burrowers, climbers and jumpers. Together with certain rodents, bats and the monotremes, these make up the sum total of native animal life in the region.

The carnivores belong mainly to the Dasyuridae and include a wide variety of animals. The marsupial mice are all small creatures similar to shrews and rats. There are different forms living in forests, deserts and even in domestic buildings. The jerboa marsupials are in every way like the well-known desert jerboas. The 'native cats', are spotted and blotched carnivores of

▼ Common opossum (*Didelphis marsupialis*)

▲ Western native cats (*Dasyurus geoffroyi*)

the genus *Dasyurus* and related species. Among these are the Tasmanian devil (*Sarcophilus harrisii*), looking rather like a bear cub, the thylacine or Tasmanian marsupial wolf (*Thylacinus cynocephalus*), today almost extinct thanks to human ravages, and the numbat (*Myrmecobius fasciatus*), a handsome forest anteater. The Notoryctidae or marsupial moles, are also found in this region, but will be described later when dealing with animals that live underground.

Other marsupial insectivores or carnivores inhabiting the Australasian woods and forests are the bandicoots of the family Peramelidae, some of which excavate holes similar to those of rabbits. There are numerous arboreal species, with prehensile tails, belonging to the Phalangeridae. Some of the latter are also to

▲ Short-headed glider (*Petaurus breviceps*)

be found in the more northerly islands as far away as Celebes. They vary in size, some being as tiny as mice, some as large as cats. All are covered in soft, woolly fur and most of them are nocturnal plant-eaters. Among them are species similar to the flying squirrels, such as the animals of the genera *Acrobates* and *Petaurus*. They are commonly called flying phalangers or gliders, and perform gliding leaps from tree to tree. The cuscuses (*Phalanger*) resemble monkeys with beautiful plain or spotted fur. They are slow-moving, solitary creatures, typical of the rain forests of New Guinea and eastern Australia. The few species of possums belonging to the genus *Trichosurus* include animals that live happily in parks and under the eaves of houses.

▼ Spotted cuscus
(*Phalanger maculatus*)

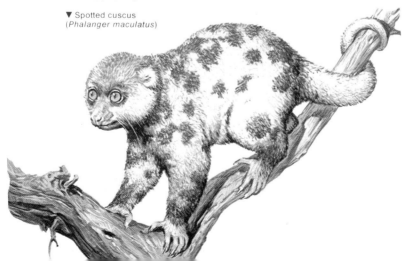

▲ Banded anteater or numbat (*Myrmecobius fasciatus*)

Probably the most famous woodland marsupial is the koala (*Phascolarctos cinereus*). This tree-dwelling animal feeds exclusively on eucalyptus leaves. It has a rudimentary tail and a pouch which, unlike that of most other marsupials, opens to the rear. The strange-looking wombats (Vombatidae) are creatures both of forest and savanna. The several species are powerful, stocky animals with nocturnal habits. They are given to burrowing, and resemble large marmots. In teeth structure and diet they are similar to rodents.

Of all the marsupials of the Australian region, the most typical are the kangaroos of the family Macropodidae. They are immediately recognisable by their huge hind legs, specialised for jumping. Because of their type of life and the correlated anatomical structure of their digestive system, the kangaroos play the same role in Australasia as the ungulates, or the ruminants, in the rest of the world.

The most unusual members of the Macropodidae are the tree kangaroos (*Dendrolagus*), living in the forests of New Guinea. They make agile leaps from tree to tree, in contrast to their ground-dwelling relatives who move about by means of a series of horizontal jumps. The ground kangaroos are to be found in every type of habitat in Australia, Tasmania, New Guinea and the Bismarck archipelago, including rain forests, savannas, plains and inland deserts. All species have a marsupium and the females generally give birth to a single baby.

Some members of the family including the rat-kangaroos, are about the size of rabbits. They usually live in woods and forests.

▲ Tree kangaroo (*Dendrolagus*)

Of these rat-kangaroos, some are either rare or even recently extinct, as a result of man's interference. They have been hunted, used as poisoned bait against introduced rabbits or fallen victim to introduced predators such as foxes. One species lives in deserts and on plains and another, the musky rat-kangaroo (*Hypsiprymnodon moschatus*) inhabits the rain forests of Queensland, feeding on plants and insects.

The true kangaroos and wallabies are all herbivores, usually content to browse on grassy plains. The smallest species of these typical kangaroos are the wallabies which often boast an elaborate striped or spotted coat. They mainly live among rocks or trees. Some wallabies are found in the dense forest zones of New Guinea. The three giant species of kangaroo belonging to the genus *Macropus* are animals of the savannas and sparse woodlands of western Australia. They are the wallaroo (*Macropus robustus*), the grey kangaroo (*M. giganteus*) and the red kangaroo (*M. rufus*). These are specialised herbivores with highly developed hind legs and a huge muscular tail. They live in large gregarious herds and move about at considerable speed with characteristic leaps. Undoubtedly these kangaroos are the most familiar inhabitants of the Australian savannas and prairies.

This brief survey can give only a rough idea of the variety of marsupial life (there are some 120 species) in the Australian region. Mention has been made, however, of the fact that the native fauna is not wholly composed of monotremes and marsupials, for there are about 60 species of rodents (rats and mice) and about 40 species of bats. All these placental mammals are fairly small. The largest is a member of the dog family, the dingo (*Canis dingo*), and is an introduced species. Thus they are neither as typical nor significant as the kangaroos, wombats, native cats and the like.

Placental mammals arrived on the Australian continent a long time after the first colonising wave of monotremes and marsupials. The rodents managed to cross the short distances of ocean between the Oriental and Australian regions by clinging to floating vegetation. They subsequently evolved, as mentioned above, into some five dozen species. The bats reached Australia without difficulty by flying from one island to another. More recent arrivals in Australia were man and the dog, who have had a strong influence on the ecology of the isolated continent. In some cases their influence has culminated in the decimation and the disappearance of the earliest colonisers.

Red kangaroo (*Macropus rufus*) with joey in pouch (marsupium) ▲

Before Australia was separated from Asia, the marsupials were the dominant mammals in Eurasia and in America, even though the placental mammals destined to supersede them were already in existence. It was the marsupials who made use of the final land links between Australia and the rest of the world. Given the striking variety of geographical and climatic zones in Australia, the invading marsupials became differentiated and eventually occupied all available ecological niches by the process of adaptive radiation. This is a typical phenomenon when a new area, containing numerous habitats, is colonised. Not only will a number of species thrive alongside one another but each species will undergo evolutionary divergence. This results in the appearance of dissimilar new species and subspecies, sometimes bearing little resemblance to the parent stock. This is how the marsupials were able to develop and adapt in parallel fashion to the various orders of placental mammals in the rest of the world, so that in due course there were marsupial moles, marsupial flying squirrels, marsupial anteaters, marsupial rat-kangaroos, marsupial wolves and so forth.

The ecological regions

The following sections examine the distribution of present-day mammals in their various biomes, which are the different climatic and vegetational belts that are to be found on the earth's surface.

The placental mammals, Eutheria, make up the vast majority of the entire class Mammalia, representing about 94 per cent of known species. They populate virtually all parts of the earth and include two unique species of indigenous mammals in New Zealand, the so-called 'land without mammals'. In Australia itself, the kingdom of the marsupials, there are, as previously said, about 100 species of placental mammals. The distribution of about 4,000 placental species is extremely varied. In discussing the zoogeographical regions, mention was made of the fact that some orders of mammals are endemic to particular regions. Although this is a rare phenomenon in the case of a large group, it more or less tends to be the rule for species, genera and sometimes families. Furthermore, many species only live in particular habitats within their area of distribution, while other species may live in a number of different habitats.

The largest number of species live on the ground. This is where the most primitive mammals, as well as their synapsid reptile ancestors, lived without any specialised appendages. However, when mammals began to evolve at the end of the Mesozoic and in the Tertiary, there is evidence of the appearance of a variety of specialised structures. These enabled mammals to adapt to all the varied habitats on the earth's surface. In earlier chapters reference has been made to the differences in shape, structure, methods of locomotion and life patterns which enabled placental mammals to thrive in a wide variety of environments. What happened was a large-scale process of adaptive radiation, similar to that which occurred, to a lesser extent, among the marsupials of Australia. The phenomenon can also be seen in the most primitive order of all, the insectivores, who are regarded as resembling the ancestors of all other eutherians. This order has many characteristics of later groups of placental mammals, including an allantoic placenta (making it possible for the embryo to remain longer inside the mother's womb), the absence of a marsupium and epipubic bones, a corpus callosum linking the cerebral hemispheres, and a dentition made up of 44 teeth, differentiated and generally reduced in number when the permanent set appears. Among these insectivores it is possible to distinguish various standard forms. Thus the 'shrew' type comprises animals that are usually small, with a long muzzle, short, five-toed feet and a fairly long tail. Most are plantigrades, scurrying through the dense undergrowth and feeding on tiny invertebrates. Although there is nothing unusual about their outward appearance, their teeth are fairly specialised for catching insects, with large incisors forming pincers for gripping prey. Some are even more specialised in being adapted for aquatic life. Thus the African otter shrews (*Potamogale*) as their name suggests, resemble small otters.

The 'mole' type comprises species that have undergone more noticeable modifications for subterranean life. The 'hedgehog' type is represented by larger insectivores such as hedgehogs themselves (Erinaceidae) and tenrecs (Tenrecidae). Their bodies are equipped with sharp, defensive spines. Finally there is the 'jumping' type, consisting of the elephant or jumping shrews (Macroscelididae). This insectivore family possess highly developed hind legs for leaping, and look rather like rat-kangaroos.

Even within a single order, comprising small, primitive and generally homogeneous species, one finds differences in the various types of habitat where they live. Among other orders, however, there tends to be a greater measure of homogeneity and specialisation in the group as a whole, but much more striking variation between the different orders. The ungulates, for example, are all adapted for running across prairies and savannas, although some specialised groups are suited for life in high mountain regions or dense forest zones. The carnivores are all specialised predators, but show notable differences according to the type of habitat where they hunt. Primates are typically

arboreal but some species are suited for life in dense forests and
rocky regions. Thus it is convenient to look at these placental
mammals, both in relation to their distribution around the world
and their particular habitats, starting with the forests.

The forests

The various types of vegetation making up the forests go through
a succession of growth and development phases before attaining a
state of relative equilibrium. They adapt themselves to different
physical situations that are closely linked to factors such as the
nature of the substratum and the climate. Looking at the land
areas from north to south, it is apparent that there is a difference
in the appearance and composition of the belts of forest that
succeed and often overlap one another. These vary according to
the changes in climate. The most northerly forests make up a
broad belt of the Holarctic region immediately to the south of the
tundra. First comes the taiga, composed principally of evergreen
conifers. In the more temperate zones farther south there are the
various types of broad-leaved deciduous forest. Finally, in the
tropical areas, there are evergreen, broad-leaved forests (tropical
rain forests) in regions with a high incidence of mainly non-
seasonal rainfall, or broad-leaved deciduous forests in areas that
have a regular dry season.

The taiga

The immense belt of coniferous forest covering the whole of
northern Siberia, Russia, Scandinavia and Canada, is called the
taiga. Its extremely low temperature varies according to the
distance from the sea. Some of the mammals inhabiting the taiga
are to be found there all the year round, some seek refuge there
only in winter or summer. Most of their adaptations are
associated with the low temperatures and the thick snow that
blankets the ground. Thus the specialised foot structure of the
Arctic hare has the effect of broadening the surface of support on
snowy terrain and preventing the animal from sinking.

As in other forest biomes, so there are various levels of the taiga
itself, all occupied by different mammals. Insectivores and small
rodents confine their activities to the detritus on the ground. Such
animals include certain red-backed voles (*Clethrionomys*) and the
wood lemming (*Myopus schisticolor*). Larger ground mammals
comprise both herbivores and predators. Up in the trees there are
rodents (squirrels and flying squirrels) and carnivores (martens).

The many herbivorous mammals of the taiga include all the
rodents and ungulates. The rodents include species living up in the
branches which feed on leaves, buds and fruits, such as the red

▲ Giraffe (*Giraffa camelopardalis*)

119

squirrel (*Sciurus vulgaris*), often found here all the year round. Then there are the graceful striped squirrels known as chipmunks (the genus *Eutamias* in America and Siberia, and the genera *Tamias* and *Tamiascurus* in North America), as well as the flying squirrels represented by *Pteromys volans* in Siberia and *Glaucomys volans* in America. The lagomorphs are represented by the snow or varying hare (*Lepus timidus*). The ungulates include the Cervidae as well as one member of the Bovidae, the Siberian ibex (*Capra sibirica*).

The deer are the largest mammals of the taiga. The giant of the family is the elk or moose (*Alces alces*). It has enormous palmate antlers, and inhabits both Eurasia and North America. Other Cervidae living in the taiga – either with a widespread distribution through the forests of the Holarctic region or in the Palearctic region only – are the red deer (*Cervus elaphus*) of Europe and Asia, its North American subspecies, the wapiti (*Cervus elaphus canadensis*), the roe deer (*Capreolus capreolus*) and the musk deer (*Moschus moschiferus*) of the Siberian taiga. In winter there arc also enormous herds of reindeer (*Rangifer tarandus*). This species has a wide distribution throughout the Arctic regions of Eurasia and North America, where it is called the caribou.

▼ Elk (*Alces alces*)

Among predators that live only in the taiga are mustelids, such as the wolverine or glutton and some martens. The most famous marten is the sable (*Martes zibellina*), which is renowned for its magnificent fur. A typical inhabitant of Siberia, the sable is a skilful hunter of mice, squirrels, flying squirrels and also large birds such as grouse. For the sable, as for other mustelids, insects and berries also form an important part of the diet. The same type of life is led by the North American fisher or pekan (*Martes pennanti*). In Europe the pine marten (*Martes martes*) is found in more southerly regions of forest down to the Mediterranean. The glutton or wolverine (*Gulo gulo*), largest of all land mustelids, is exclusive to the taiga, and although an able climber usually hunts on the ground. This species also roams all over the Holarctic region, although not in large numbers. The animal is very rare in Europe. It is an extremely efficient predator, capable of killing fairly large animals, particularly in winter when snow makes it easier to follow the tracks of prey. During this season it also captures reindeer and young elk. The glutton also feeds on carrion and often steals the remains of prey left by bears, wolves and lynxes. It can therefore be considered a superpredator. Although it has virtually no enemies, apart from man, an adult glutton may

▼ Pine marten (*Martes martes*)

occasionally be attacked by a wolf. The young sometimes fall victim to bears, wolves and foxes. In spring, when the woods and forest are filled with large flocks of migrating birds that nest there, the glutton feeds mainly on fledglings. In summer its food consists chiefly of insects, berries and small rodents.

Other carnivores of the taiga, though not exclusive to these parts, are mustelids, such as weasels (*Mustela nivalis*) and minks (*Mustela lutreola* in Eurasia, *M. vison* in America), together with lynxes, foxes, wolves and brown bears. The wolf (*Canis lupus*) is much more frequently seen in open regions like the tundra, while the brown bear (*Ursus arctos*), particularly in the Holarctic region, more often roams other forest zones. The Himalayan or Asian black bear (*Selenarctos thibetanus*) is occasionally found in some eastern parts of the taiga. Its counterpart in the New World is the American black bear (*Ursus americanus*). The red or common fox (*Vulpes vulpes*) is one of the most familiar species of the taiga, but this animal also has an extremely wide range throughout the Palearctic and, to some extent, Oriental regions. The North American populations are regarded by some authors as subspecies of the common fox. Others, however, regard it as a closely related species in its own right (*Vulpes fulva*). In North America there are a variety of foxes which have different shades of fur. One of the most famous is the silver fox.

▼ Red or common fox (*Vulpes vulpes*)

In the eastern taiga the raccoon-like dog (*Nyctereutes procyonoides*) is the only member of the dog family to spend the winter hibernating. Some cats are also found in the taiga although their habitat is not limited to this region. The European lynx (*Lynx lynx*) is a formidable predator, hunting large herbivores. The tiger (*Panthera tigris*) still roams the forests of eastern Siberia. These Siberian tigers, largest members of their species, are the most powerful and efficient predators to be found in the coniferous forests which, for most of the year, are cloaked in snow. Their prey includes red deer, elks, musk deer, roe deer and wild boars.

▼ European red squirrel (*Sciurus vulgaris*)

The temperate forest

To the south of the taiga lies an enormous temperate zone,
consisting of central Europe, part of eastern Asia and the United
States. There are many different habitats in central Asia. These
are separated by long chains of high mountains and deserts. In the
southern hemisphere there is an equivalent temperate belt running

▼ Common or European lynx (*Lynx lynx*)

through Australia, New Zealand and South America. The
essential feature of both these zones is a notable difference in
climatic and atmospheric conditions from season to season.

The typical forest of the temperate zone consists largely of
broad-leaved deciduous trees. The flowers, plants and trees of the
forest vary considerably from one region to another, although

Tiger (*Panthera tigris*) ▶

basically there are analogies between the flora of Eurasia and North America. The forest climate is subject to fewer and less sudden changes of temperature than the more open habitats (such as prairie and cultivated land) in the same temperate zone. The trees of the forest grow very close to each other and thus create an effective windbreak, retain a high measure of moisture and provide warmth and shelter for many animals. Nevertheless, conditions within the forest vary quite considerably with the changing seasons. In winter, when the trees are stripped of leaves, the temperature may be as cold as on the Arctic steppes. During summer it may be as hot as in the tropical rain forest. The animal inhabitants of the temperate forests therefore have to be able to adapt to climatic variations.

Today deciduous forests have been reduced by man to a few isolated fragments. In Europe entire forests were savagely and thoughtlessly destroyed to clear land for farming. It is only recently that the science of silviculture has been developed. The appearance of these residues of temperate forest varies noticeably according to climatic and local geographic conditions, exhibiting all types of intermediate, transitional forms between the extremes of taiga and subtropical forest.

The large amount of vegetation produced in these broad-leaved temperate forests attracts many herbivorous mammals from large ungulates (Cervidae) to small rodents. Preying on this community of herbivores are the various carnivores. Some large mammals, such as the bear and the wild boar, get the best of both worlds, for they are omnivores, and feed either on plant matter or on other animals, depending on season and place. The brown bear, already encountered in the taiga, finds conditions in the temperate and mountain forests ideal. It has a very wide geographical range embracing the whole region, but there are a number of well differentiated populations, now recognised as being subspecies of the single species *Ursus arctos*. Thus in Eurasia average sizes tend to be fairly modest. The European brown bear, now confined to the Apennines and the Pyrenees, weighs up to 250 kg while the Russian forms may exceed 300 kg. The grizzly bear (*Ursus horribilis*) can weigh as much as 360 kg. It is also now rare and restricted to mountain forests. The gigantic Kodiak bear of Alaska can weigh up to 780 kg. No matter where it is found, however, this omnivorous species is capable of exploiting the food resources of a range of diverse habitats. In summer its diet consists basically of vegetable matter, including fruits, berries, roots and fungi. In autumn its diet consists of acorns, chestnuts and beechmast. In the spring this diet is supplemented by small animals including insects caught beneath stones and in rotten tree trunks, insectivores and rodents, young wild boars, sick or injured chamois and roe deer, and even domestic animals. The bear also catches birds and fishes, and will vary its diet with eggs, honey, frogs and snails. It is essentially a nocturnal creature. The males are solitary animals but the females live in small family groups.

The other dominant mammal in temperate forests is the wild boar (*Sus scrofa*). It is also an omnivore with mainly nocturnal habits. The species has a wide range in the Palearctic region but spreads south as far as Sudan, India and other parts of the Oriental region. In addition to feeding on berries, acorns and chestnuts, the wild boar is a specialised digger. With the aid of its muzzle which ends in a flat, rough disc, it rummages in the ground for roots, tubers, truffles and small animals. It can also catch rabbits, small rodents and fishes. The rooting habits of the wild boar, resulting in deep furrows in the ground, are important factors in the woodland ecology, for this natural ploughing activity creates good conditions for the germination and growth of new trees.

▼ Brown bear (*Ursus arctos*)

The specialised herbivores of the temperate forest are the red deer and roe deer in Europe and Asia. There are also a few other species of Cervidae from eastern Asia, such as the sika (*Cervus nippon*). In North America there is the splendid wapiti, already mentioned in relation to the taiga, but which, like the moose, roams the deciduous forests. In the Mediterranean region a familiar species is the fallow deer. Among the Bovidae, the European bison or wisent (*Bison bonasus*) is extremely rare, and now confined to certain parts of Poland and Russia.

Among a large number of tiny rodents are woodmice (*Apodemus*), red-backed voles (*Clethrionomys*) and representatives of the Gliridae such as the edible or fat dormouse (*Glis glis*), the hazel dormouse (*Muscardinus avellanarius*) the garden dormouse (*Eliomys quercinus*), and the squirrels, both red (*Sciurus vulgaris*) and grey (*Sciurus carolinensis*).

Typical carnivores of the temperate forests are the lynxes. The Mediterranean species is the Spanish lynx (*Lynx pardina*). The very beautiful wild cat (*Felis sylvestris*) is a European species that once had a wide distribution but is now very rare and confined to the Scottish highlands and woodlands in southern and eastern Europe. It is a solitary, fierce predator. The wolf is a superpredator of the temperate forests, although today its numbers are much reduced.

▼ European
wild boar (*Sus scrofa*)

Red deer (*Cervus elaphus*) ▶

◀ European wild cat
(*Felis sylvestris*)

▼ Eurasian badger (*Meles meles*)

Mustelids found in these forests include the badgers (*Meles meles* in Europe, *Taxidea taxus* in North America), and the pine marten (*Martes martes*), which is adept at climbing trees. The New World is also the home of all the black-and-white skunks, notably the striped skunk (*Mephitis mephitis*) and the spotted skunk (*Spilogale putorius*). Other Eurasian mustelids are the polecat (*Mustela putorius*), the beech marten (*Martes foina*) and the weasel (*Mustela nivalis*).

The tropical and equatorial forest

Forests are the most typical geographic feature of the immense tropical belt situated to the north and south of the equator. They are found in central and South America (the Neotropical region), central Africa (the Ethiopian region) and southern Asia (the Oriental region).

The tropics provide ideal growing conditions for forests with trees of considerable size, and the temperature is sufficiently high for such growth to be steady and continuous. Annual rainfall normally exceeds 1.5–2 m and is distributed fairly regularly throughout the year. The temperature varies only a little from season to season, and there is little difference between day and night temperatures. In the equatorial zone it averages about 25°C.

▼ Fieldmouse (*Apodemus*) with young

131

Because of heavy cloud, evaporation is very low and so the humidity tends to be exceptionally high. The tropical rain forest is influenced by natural forces that preserve a delicate state of equilibrium. The slightest modifications in climate may be enough to cause these forests to retreat or advance.

The forest is distinctly divided into layers or levels. The tall trees offer a large number of ecological niches that are either absent or negligible in other habitats. The tallest trees may grow to 40–50 m and because so little light can filter through the dense foliage there is no possibility of much grass growing below. Moreover, since the tree trunks are so tightly packed together and the network of branches and leaves so tangled, the range of vision is limited.

In Africa the area of land covered by forests is not all that large in relation to the size of the whole continent. Nevertheless, the African forests are quite extensive, with clear boundaries separating them from the adjacent savannas. An almost continuous forest belt stretches from the Gulf of Guinea to the Ruwenzori Mountains. Within this area are many interesting mammals. Some of them are shy and extremely difficult to observe. The okapi, the giant forest hog, the forest antelopes and the pygmy hippopotamus have only recently been discovered by scientists. Certain species are endemic to the African rain forests, and these deserve special mention. Among the primates there are the potto (*Periodicticus potto*) and the bushbabies (*Galago*), so named because of their strange calls, like the crying of babies. These are all lemur-like mammals with nocturnal habits, living in the trees. The Old World monkeys belong to the family Cercopithecidae and the many species of guenons (*Cercopithecus*) are more or less localised at certain levels of the forest. They have a complex social organisation and live in noisy bands high up in the dense foliage. The splendid colobus monkeys or guerezas (*Colobus*) are remarkable acrobats. The larger mandrill (*Papio sphinx*) and drill (*Papio leucophaeus*) are notable because the males display naked patches of bright colour. Best known of all these primates are the chimpanzee (*Pan troglodytes*) and the gorilla (*Gorilla gorilla*). This is the giant of the anthropoid apes, inhabiting lowland and mountain forest regions of central Africa.

Exclusive insectivores of the African forests are the pangolins, comprising four species of the genus *Manis*. Two of the species, *M. gigantea* and *M. Temmincki* being terrestrial, and two, *M. tricuspis* and *M. tetradactyla*, are tree-dwellers. All are covered in scales and feed wholly on insects such as ants and termites which they catch with their long, sticky tongues.

▲ Gorillas (*Gorilla gorilla*)

▲ Leopard (*Panthera pardus*)

▼ Black-bellied pangolin (*Manis longicaudatus*)

Typical carnivores roaming these forests are felines such as the golden cat (*Felis aurata*) and the handsome leopard (*Panthera pardus*). The leopard is not an exclusive forest dweller since it also frequents savannas and deserts. The leopard is a skilful hunter, its prey including large ungulates, monkeys, rodents, birds and reptiles. Smaller carnivores are members of the family Viverridae, such as the various genets (*Genetta pardina, G. servalina, G. victoriae, G. villiersi*, etc), the African linsang (*Poiana richardsoni*) and the African tree civet (*Nandinia binotata*).

Forest herbivores include a number of small antelopes, such as the duikers (*Cephalophus*). They live on their own or in compact groups, feeding on the shoots and leaves of low branches. The small royal or dwarf antelope (*Neotragus pygmaeus*) is also found here. The largest forest antelope is the bongo (*Taurotragus euryceros*), a nocturnal creature living alone or in small groups. The coat is reddish-brown with vertical white stripes on back and flanks. The muzzle is black and there are white patches between the eyes, on the cheeks and on the throat. The animal is well camouflaged in its forest environment of shifting light and shade. The other large forest ruminant is the okapi (*Okapia johnstoni*), a member of the giraffe family. This is a rare animal of solitary, nocturnal habits, with black and white striped haunches and legs. It also has a long, prehensile tongue. This species, only discovered in 1901, is restricted to the almost impenetrable forest regions of Zaire and Uganda.

Two species of swine (Suidae) inhabiting the African forests are the omnivorous bushpig or river hog (*Potamochoerus porcus*), which is widely distributed in equatorial Africa, and the giant forest hog (*Hylochoerus meinertzhageni*). The large forest hog, weighing up to 300 kg, was only discovered in 1904 and its habits

are still largely unknown. In the swampy forests of the Ivory Coast, Sierra Leone and Liberia, there is another strange and little known species, the pygmy hippopotamus (*Choeropsis liberiensis*). It is much less fond of water than its larger relative and is a rather solitary creature. It stands only 80 cm high and weighs not more than 250 kg.

The buffalo and the elephant have also settled in the equatorial forest although these populations differ very strikingly from those of the savanna. The forest elephant (*Loxodonta africana cyclotis*) is much smaller than its savanna relative. It has small, rounded ears and straight, downward-pointing tusks. These are clearly an advantage when moving about among closely packed trees. The dwarf or forest buffalo (*Syncerus caffer nanus*) is also relatively small and light, with shorter horns that point to the rear and hair that may be reddish-brown or black. This buffalo is an inhabitant of the dense western forests.

The mammals that live on the island of Madagascar make up a very specialised community, for most of the typical African

▼ Linsang (*Poiana richardsoni*)

▼ Pygmy hippopotamus (*Choeropsis liberiensis*)

species are not found on the island. The exceptions include certain rodents, some shrews and the bushpig, which were possibly introduced from the mainland. Because of their mobility, however, only one genus of bat is indigenous to the island. The other dozen genera are also found in Africa and in the Oriental region. However, the rest of the fauna is exclusive to Madagascar and has been so for many millions of years, probably dating back to a period before the beginning of the Tertiary.

The insectivores are represented by an indigenous family, the Tenrecidae, comprising some 30 well diversified species. This is another classic instance of adaptive radiation, for the tenrecs bear a strong outward resemblance to shrews, water shrews, moles and hedgehogs.

The most noteworthy group of mammals on the island of Madagascar is that comprising the lemurs (Lemuridae). These are primitive forms of primates, varying in size and shape. Some of them are very rare and on the verge of extinction. One of the rarest and largest is the indri (*Indri indri*). It is a tree dweller,

▲ Okapi (*Okapia johnstoni*)

▲ Ring-tailed lemurs (*Lemur catta*)

coloured black and white like a giant panda, and lives in small groups. Another threatened species is the aye-aye (*Daubentonia madagascariensis*), which feeds mainly on insects embedded in trees. The aye-aye manages to extract the larvae from their tunnels, with the third finger of each hand, which is as long and thin as a stick. The smallest of these species is the lesser mouse lemur (*Microcebus murinus*). Best known, however, are the many species of true lemurs such as the ring-tailed lemur (*Lemur catta*), which has black and white rings on its tail.

Indigenous carnivores of Madagascar belong to the Viverridae, with six genera. They include ground animals and tree dwellers, with diurnal and nocturnal habits, some omnivorous, others

▲ Lesser mouse lemur (*Microcebus murinus*)

carnivorous. All are very rare and little is known about them. The fossa (*Cryptoprocta ferox*) is as large as a fox. This rare carnivore hunts by night, preying on lemurs in the branches of trees.

The Asiatic rain forest is particularly extensive in south-eastern Asia and Malaysia. It is populated by several mammal species that may be regarded as counterparts of species already encountered in Africa and South America. There are, for example, the fruit-eating bats of the genus *Pteropus*. These include the kalong of Malaysia (*Pteropus vampyrus*) which is the largest bat in the world. These bats are also known as flying foxes. There are various endemic insectivores, about which little is known except the generic names, *Echinosorex* and *Hylomys*.

Primates are fairly numerous in the Asiatic forests, and include the primitive, indigenous family of Tupaiidae. These tiny tree shrews are all rapid runners and climbers. The Lorisidae, belonging to the suborder Prosimii, resemble the African galagos. Among them are the slender loris (*Loris tardigradus*) and the slow loris (*Nycticebus coucang*). Another small group of endemic Asiatic prosimians is the family Tarsiidae. The tarsiers are acrobatic little creatures with enormous eyes and unusual fingers terminating in suckers. There are only three species of the single genus *Tarsius*, the best known being the eastern tarsier (*Tarsius spectrum*) of Celebes.

Despite being arboreal the gibbons of the subfamily Hylobatinae are the only monkeys that spontaneously adopt the bipedal stance when on the ground. Their walk is, however, rather odd and clumsy. The gibbons are also noteworthy for their exceptionally long arms and legs, especially the former, and for the absence of a tail. Five species of the genus *Hylobates* are known, with a distribution extending to Sumatra and Borneo. Larger than these is the siamang (*Symphalangus syndactylus*). This inhabitant of the mountain forests of Sumatra, lives at an altitude of almost 3,000 metres. The siamang has a conspicuous red sac-like appendage on the front of its neck and when it utters a cry this sac swells up, contrasting vividly with the black of the fur.

The subfamily Ponginae, which contains the African species of chimpanzee and gorilla, is represented in the forests of Sumatra and Borneo by the orang-utan (*Pongo pygmaeus*), the common name being the Malayan term for 'man of the woods'. This huge tree-dwelling ape lives alone or in pairs. It feeds on fruit, leaves and small animals. The orang-utan's survival has been threatened by man's destruction of its forest environment and by the capture of many animals for zoos. Over the last century the orang-utan population has been reduced to 3,000–4,000. Virtually all of these animals live in the Borneo forests. Typical of the Cercopithecidae in this region and habitat are the macaques, the majority of which have adapted to life at ground level. Among these are the lion-tailed macaque (*Macaca silenus*) from India, the toque monkey (*Macaca sinica*) of Ceylon and the Malayan crab-eater (*Macaca irus*). Their distribution extends from Burma to Timor which is the extreme eastern limit of primate diffusion. The last species lives mainly along the banks of rivers and seas, feeding on crustaceans and molluscs. In mountain forests (including the Himalayas) the most familiar and widely distributed species is the common or rhesus macaque (*Macaca mulatta*). This monkey is famed for its experimental use in laboratories. Another well known species is the Japanese macaque (*Macaca fuscata*).

▲ Orang-utan (*Pongo pygmaeus*)

Other tree-dwelling monkeys from Asia are the langurs or leaf monkeys of the genera *Presbytis*, *Pygathrix* and *Nasalis*. The hanuman or entellus monkey (*Presbytis entellus*) is famous as the sacred monkey of India. Some races are conditioned to life among the snow-capped mountains at heights of up to 4,000 metres. Other langurs are the dusky leaf monkey (*Presbytis obscurus*), an inhabitant of the Malaysian forests, and the proboscis monkey (*Nasalis larvatus*). This animal has a curious drooping nose, which is particularly large in older males.

The most characteristic rodents of of the Asiatic forests are the various squirrels of the genera *Funambulus*, *Callosciurus* and *Ratufa*. The latter are giant, nocturnal squirrels up to one metre in length. They are strikingly coloured, and include eggs and young animals in their diet.

▲ Toque monkey (*Macaca sinica*)

Carnivores contain members of both the Mustelidae and Viverridae families. Among the Viverridae are the linsangs (*Prionodon*) and the Asiatic civets. The large Indian civet (*Viverra zibetha*) with thick spotted fur, is bred for the production of commercial civet, which is secreted by the perineal glands, and is used in perfumery. The largest of the Viverridae is the binturong (*Arctictis binturong*), which clambers through the trees like a bear cub. It has thick blackish fur and hunts by night. There are also various mongooses such as the Indian grey mongoose (*Herpestes edwardsi*), with striped fur, renowned for its ability to kill snakes. Other notable carnivores are the ferret badgers of the genus *Melogale*. These comprise several species of small badgers from China, Java and Borneo.

Asiatic feline include the marbled cat (*Felis marmorata*), which is one of a number of wild forest cats, as well as tigers and leopards. The tiger, although not confined to this habitat, is quite at home here. The inhabitants of the tropical rain forests are smaller than the massive tigers of the northern forests, and their numbers are seriously threatened. The leopard (*Panthera pardus*) is more frequently seen and has a wide range of distribution in the Asiatic forest as well as on open ground and in mountain woodlands. In addition to the familiar spotted form, there is a rarer black mutant, popularly known as the black panther. Although not as endangered as the tigers, the leopard population is also declining steadily and its range gradually being cut back. Less familiar and somewhat smaller is the clouded leopard (*Panthera nebulosa*). This indigenous inhabitant of the forests of south-eastern Asia spends much of its time in trees. The Malayan or sun bear (*Helarctos malayanus*) is another forest carnivore, and is smaller than most other bears. Among the Canidae there is the gregarious predator variously known as the dhole, red dog or Indian wild dog (*Cuon alpinus*). These creatures hunt in packs of several dozen and are capable of pulling down large deer.

The Indian or Asiatic elephant (*Elephas maximus*), often tamed as an adult, is a forest dweller and still quite numerous in Assam, Ceylon and Burma. However, the three species of Asiatic rhinoceroses, are dramatically decreasing in numbers. The Javan rhinoceroses (*Rhinoceros sondaicus*) have been reduced to only a few dozen individuals, and the number of Sumatran rhinoceros

▼ Leopard (*Panthera pardus*)

▲ Asiatic elephants (*Elephas maximus*)

(*Didermocerus sumatrensis*) is now less than one hundred. Both
species are therefore on the brink of extinction. The great Indian
rhinoceros (*Rhinoceros unicornis*), largest of the three, is not so
threatened, although confined to reserves in India and Nepal.

The Malayan tapir (*Tapirus indicus*), a perissodactyl, is the
only Asiatic species. It lives in Burma, Thailand, Malaysia and
Sumatra. Its two relatives are South American animals. Forest
artiodactyls include several species of wild boars, the most
famous being the babirusa (*Babyrousa babyrussa*). The

characteristic tusks of the male are continually growing, turning up and backwards. This animal is restricted to the islands of Celebes, Togian, Buru and Sula, frequenting rain forests and reed thickets on the banks of rivers and pools.

Representing the Cervidae in this same habitat are several attractive species of deer. The large sambar (*Cervus unicolor*) is a heavily built animal with only slightly branched antlers. The handsome axis or Indian spotted deer (*Cervus axis*) has a reddish coat with white spots. The little hog deer (*Axis porcinus*) is only

▲ Indian rhinoceros (*Rhinoceros unicornis*)

about a metre tall. Other small deer of the Oriental region are the muntjacs or barking deer (*Muntiacus*). They have short, pointed antlers which sprout from an elongated bony pedicle.

Perhaps the most marked difference between the Asiatic forest and the African forest is the distribution of Bovidae. In Africa there is only the dwarf buffalo, but in Asia there are many species of cattle, including the Indian or water buffalo (*Bubalus bubalus*). This species is now very rare in the wild but extremely common in its domesticated form, not only in India but also in mountain regions and alongside rivers throughout Eurasia. In Europe it is found in France, Italy and Hungary. There are two dwarf species of buffalo. The anoa (*Anoa depressicornis*) is found in Celebes. The males stand up to one metre high. The tamarau (*Anoa mindorensis*), is a little taller and is a native of the island of Mindoro in the Philippines. There are also three other species of wild cattle belonging to the genus *Bos* (true cattle). All have ringed horns, and are now restricted to Asia. The gaur (*Bos gaurus*) is a timid forest animal, standing up to two metres high. It often falls prey to the tiger. In the wild it is in grave danger of extinction because of intensive hunting. The domestic form is called the gayal. Just as rare is the smaller banteng (*Bos banteng*), an inhabitant of Java, Borneo and Burma. The kouprey (*Bos sauveli*) lives in Vietnam and Cambodia, although little is known about this animal.

The rain forests of the Neotropical region stretch across a vast area of Central and South America, from southern Mexico down to southern Brazil. Known as the selva, this is the largest continuous forest belt in the world. The mammal population of this habitat is very complex, and contains a large proportion of endemic groups. South America was an isolated land mass for some 60 million years, detached from North America at the

▼ Babirusa (*Babyrousa babyrussa*)

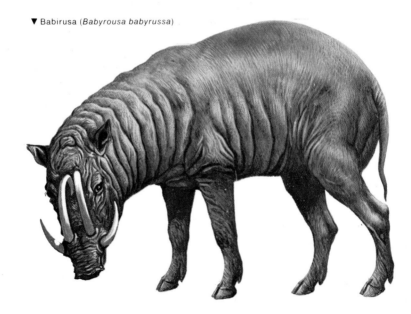

▲ Gaur (*Bos gaurus*)

▼ Binturong *(Arctictis binturong)*

147

beginning of the Tertiary. Only during the late Pliocene were the two continents reunited. This resulted in a wave of mammals from North America, which had always been linked with Eurasia and Africa. This mammal community superseded the fauna that had evolved in South America from the earliest primitive colonisers. Nevertheless, a number of ancient animal groups have survived to this day, having existed since the end of the Tertiary. The marsupials, for example, are represented by two endemic families, the Didelphidae and the Caeonolestidae (oppossums and opossum-rats). The edentates native to the region include anteaters, sloths and armadillos.

The South American forest insectivores are represented only by a few species of Soricidae (shrews) of the one genus *Cryptotis*, found in the north-western part of the continent, in Colombia, Ecuador and Venezuela. Their place is taken in the selva by small marsupials and the more specialised edentates. Among the latter, three anteaters have the same role and possess the same adaptations as the previously mentioned anteaters belonging to other orders (the marsupial numbat in Australia, the African aardvark and the pangolins of Africa and Asia). This is an especially striking and successful example of convergent evolution. The giant anteater (*Myrmecophaga tridactyla*) hunts on the floor of the forest but is also found in more open habitats. The tamandua or collared anteater (*Tamandua tetradactyla*) and the two-toed anteater (*Cyclopes dactylus*) are exclusively arboreal, and have prehensile tails.

The most characteristic inhabitants of the Neotropical forest are the monkeys, all of which are tree-dwellers. They belong to two families, the Callithricidae (Hapalidae) and the Cebidae,

▼ Giant anteater (*Myrmecophaga tridactyla*)

Common squirrel monkeys (*Saimiri sciurea*) ▶

▲ Capuchin monkey (*Cebus capucinus*)

which comprise the suborder Platyrrhini. The former family comprises about 36 species of small monkeys with thick fur and a non-prehensile tail. They are known as marmosets (*Callithrix*) and tamarins (*Leontocebus*), and have manes, ear tufts or long moustaches. The Cebidae are more varied in appearance, and occur in larger numbers. There are some 40 species of varying dimensions, some with a prehensile tail. They include more fruit in their diet than the marmosets and tamarins. One member of the family is the douroucouli (*Aotes trivirgatus*), which is also known as the night monkey because it is the only true monkey with nocturnal habits. It has huge eyes resembling those of the Asiatic tarsiers. Other Cebidae are the sakis (*Pithecia* and *Chiropotes*), the little squirrel monkeys (*Saimiri*) and the uakaris (*Cacajao*), with red, hairless faces. The capuchins (*Cebus*) are noted for their social behaviour.

The howler monkeys (*Alouatta*) are large, strongly built creatures with thick hair and long prehensile tails. Their distinctive calls play an important role in group behaviour and demarcation of territory. The voice is very powerful and deep as a result of a particular bony structure derived from the transformed hyoid bone, which acts as a sound-box. Other characteristic and wholly arboreal species are the spider monkeys (*Ateles*) and the woolly monkeys (*Lagothrix*). In both forms the limbs, especially the arms, are extremely long. They perform acrobatic feats among the branches that are comparable to those of the Asiatic gibbons, sometimes making leaps of 10 metres.

▲ Spider monkey (*Ateles variegatus*)

The Bradypodidae, of the order Edentata (commonly known as sloths), are curious animals of the South American forests. They are entirely arboreal but with a behaviour pattern in extreme contrast to that of the platyrrhine monkeys. There are six species belonging to two genera, *Bradypus* (ais or three-toed sloths) and *Choloepus* (two-toed sloths). All look rather odd, with a round head and forward-directed eyes, a rudimentary tail and exceptionally long limbs. The arms are especially long. The

▲ Three-toed sloth (*Bradypus tridactylus*)

fingers are joined together but furnished with enormous curving claws. These enable the animals to cling upside down to branches and to move very slowly along them. Sloths feed entirely on leaves and exhibit strange, specialised features that accord to their way of life. Thus the arrangement of the hair on body and limbs points in the opposite direction to that of other mammals. This feature enables rain water to run off the sloth's body as it hangs upside down from a branch. The presence of algae in the hair sometimes gives the coat a greenish colour and this provides a measure of camouflage to the sloth as it hangs motionless in the wet foliage.

The rodents of the Neotropical forest include a few species of squirrels (Sciuridae), which have a world-wide distribution. Other species indigenous to South America are cavies, capybaras, pacas, chinchillas, pacaranas, coypus, agoutis, spiny tree rats and

▲ Prehensile-tailed porcupine (*Coendou prehensilis*)

many others. Some of these rodents live in trees, as for example, the twenty or so species of porcupines belonging to the genus *Coendou*, which are exclusive to South America. These strange tree porcupines have quills and possess a long prehensile tail. Other rodents are ground dwellers and it is interesting to note how a number of the larger rodents have come to occupy the ecological niches of various small antelopes and other ungulates in Africa. They also show a surprising resemblance to these

▼ Capybara (*Hydrochoerus hydrochaeris*)

▼ Jaguars (*Panthera onca*)

African animals in respect of size, type of life and even coat coloration. Thus the agoutis (*Dasyprocta*) correspond to the royal antelopes (*Neotragus*). The paca (*Cuniculus paca*) is very similar to the only member of the African Tragulidae, the water chevrotain (*Hyemoschus aquaticus*), both in general shape and the distinctive markings on the back.

The few species of Cervidae include the Virginian or white-tailed deer (*Odocoileus virginianus*), this being the southernmost limit of its broad range, and the marsh deer (Odocoileus dichotomus), rare and little known. The most typical small deer of the forest are the brockets (*Mazama*) which are the counterparts of the African duikers. Among other ungulates, the most numerous are the peccaries (*Tayassu*), small aggressive wild boars that live in herds. The perissodactyls include the two species of tapir (*Tapirus*), which are shy and solitary animals.

The carnivores are represented by two large cats, the jaguar (*Panthera onca*) and the puma (*Felis concolor*), and two smaller ones, the ocelot (*Felis pardalis*) and the jaguarondi, eyra or otter cat (*Felis eyra*). Among the Canidae, the bush-dog (*Speothos venaticus*) hunts in small packs, preying on the large rodents. Finally, there are numerous Procyonidae, such as the kinkajous (*Potos*), the coatis (*Nasua*) and the olingos (*Bassaricyon*), all of which are arboreal, as well as some Mustelidae.

▲ Collared peccary (*Tayassu tajacu*)

▲ American bison (*Bison bison*)

The temperate grasslands

In places where the amount of rainfall is lower than that of the forests but higher than that of the deserts, the predominant forms of vegetation are herbaceous plants. In tropical zones, where there are distinct dry and rainy seasons, these regions are called savannas. In temperate zones these regions are described as temperate grasslands or prairies. These grasslands cover an enormous area of Eurasia and North America where they occupy about one-third of the United States.

The most characteristic mammals of these grasslands are various large herbives. In Europe and Asia the large ungulates of the grasslands have almost completely vanished. The Mongolian wild horse or Przewalski's horse (*Equus przewalskii*) is today restricted to a very small zone of central Asia in a mountainous, semi-desert habitat. There is also a prairie species of deer from eastern Asia known as Père David's deer (*Elaphus davidianus*), which is no longer found in the wild. The Chinese water deer (*Hydropotes inermis*), a small animal without antlers from the humid plains of China and Korea, still survives in its natural

156

▲ Pronghorn antelope (*Antilocapra americana*)

habitat. However, the tarpan (*Equus gmelini*) of southern Russia, thought to be the ancestor of the domestic horse, is quite extinct. In areas where open plains have been converted into pastures, the wild herbivores have largely been replaced by domestic species. Three examples from different regions are the bison, or buffalos, of the temperate North American prairie; the many antelopes of the African savannas; and the giant kangaroos of the Australian plains. All these natural inhabitants of the grasslands have largely given way to cattle, sheep and goats.

The prairies are for the most part flat and open and annual rainfall is low. Trees are rare, offering large wild mammals little in the way of shelter. As a result some species such as the red deer, the roe deer and the wild boar, move about here only at night. At one time enormous herds of bison (*Bison bison*) migrated across the vast plains of North America and Canada. Farther west the pronghorn antelope (*Antilocapra americana*) still survives, although again its numbers have been greatly reduced.

With the exception of certain insectivores, such as the mole, the most numerous inhabitants of the temperate grasslands are the rodents. These are species that excavate the soil, and some of them, like the mole rats (Spalacidae) are extremely well adapted to life below ground. On the Eurasian grasslands the best known species is the bobac or steppe marmot (*Marmota bobak*). This large rodent from the Ukraine and western Siberia digs complex burrows where it spends the winter hibernating. Its counterpart in Mongolia and northern China is the Mongolian marmot (*Marmota sibirica*). The much smaller and more delicate-looking ground squirrels (*Citellus*) of Eurasia and North America have similar habits to the marmots, and also hibernate in winter.

The most celebrated rodents of the North American prairies are the prairie dogs, sturdy marmot-type animals. These gregarious creatures dig elaborate networks of tunnels known as

▼ Common hare (*Lepus europaeus*)

'towns'. There are two species, the black-tailed prairie dog (*Cynomys ludovicianus*) and the white-tailed prairie dog (*Cynomys leucurus*).

The carnivores of these habitats include the fox, which is found throughout the whole Holarctic region, the coyote (*Canis latran*) and the skunk (*Mephitis*) which all feed mainly on rodents.

Cultivated fields

This type of habitat is basically a modification of the temperate grasslands and the mammals to be found here are those that have adapted most successfully to man's farming activities. The main inhabitants are small rodents living in underground burrows. These include the ground squirrels, meadow voles and various

▼ Black-tailed prairie dog (*Cynomys ludovicianus*)

rabbits such as the American cottontails and related species of the genus *Sylvilagus*. In Europe the most frequent inhabitant of cultivated land is the common hare (*Lepus capensis*). There are also numerous tiny harvest mice (*Micromys minutus*) which feed on cereal crops and build circular nests between the tall stalks.

The steppes

The treeless steppes of Eurasia and northern China are notable for their exceptionally dry climate. The annual rainfall does not generally exceed 400 mm and there is high evaporation. They are cold regions with temperatures averaging between 3° and 10°C according to the latitude. Vegetation consists mainly of plants adapted to arid conditions such as grasses and low bushes. The dominant mammals here are also rodents, including many

▼ Saiga (*Saiga tatarica*)

▲ Mantled ground squirrel (*Citellus lateralis*)

burrowing meadow voles of the genus *Microtus*. Other rodents
are the steppe lemmings (*Lagurus*) and a number of ground
squirrels. The most typical species of ground squirrel found in the
European steppes is the suslik (*Citellus citellus*), frequently seen
in southern Austria and Turkey. This animal constructs fairly
elaborate burrows with several tunnels and a central chamber
accommodating the nest. The suslik can also occasionally be
found among cultivated fields. It sometimes has two periods of
dormancy during the year.

The typical ruminant of this steppe habitat is the saiga (*Saiga
tatarica*), although it is also found in desert and subdesert zones.
It is a curious-looking antelope with puffy nostrils. Only the males
have antlers. At one time the saiga had a wide distribution ranging
from Poland to eastern Mongolia. Herds containing thousands of
animals roamed the steppes, particularly of Russia, in regular
seasonal migrations. In the 19th century it was the object of
intensive hunting so that at the beginning of the present century
there were only a few hundred animals left. In 1919 the Soviet
government placed a ban on hunting and initiated a series of
studies relating to the saiga's biology and ecology. Within a few
years the saiga population underwent a dramatic increase and
today it is legal to shoot up to 300,000 animals a year. The saiga is
adapted to make best possible use of the steppe's meagre food
supply and since there are very few domestic animals, it plays an
important role in the lives of the local people.

▲ Patagonian cavy or mara (*Dolichotis australis*)

The pampas

In South America an enormous area of Argentina and Uruguay is covered by semi-arid prairie, where there are no trees and where the vegetation consists chiefly of grasses. As it extends westward it becomes steadily drier, being interspersed with thorn bushes, providing pasture for huge herds of domestic cattle. The original wild ungulates are represented by a few Cervidae, the marsh deer (*Odocoileus dichotomus*), and the pampas deer (*Odocoileus bezoarticus*). Of the many rodents the best known are the Brazilian cavy (*Cavia aperea*), wild relative of the common guinea pig, and the viscacha (*Lagostomus maximus*), whose flesh is edible. The viscacha, which is somewhat larger than a rabbit, excavates complex networks of underground tunnels.

▲ Maned wolf (*Chrysocyon brachyurus*)

The giant of the rodent family is the capybara or water cavy (*Hydrochoerus hydrochaeris*), which is found near rivers and streams and also inhabits forest zones throughout South America east of the Andes. In the colder, southern regions live the maras or Patagonian cavies (*Dolichotis*). These hare-like animals have long legs adapted for leaping and running. The tuco-tucos (*Ctenomys*) are smaller rodents.

There are no insectivores of the pampas and their place is taken by several species of armadillo. Predators are represented by the hog-nosed skunks (*Conepatus*), of the family Mustelidae, by the foxes of the genus *Dusicyon* which resemble small coyotes, and by the pampas cat (*Felis pajeros*). The maned wolf (*Chrysocyon brachyurus*) has long legs and is a swift runner.

▲ Cheetah (*Acinonyx jubatus*)

The savannas

In the tropical zones immediately north and south of the equatorial regions, there are areas of grassland commonly described as savannas. The distribution of bushes and trees varies according to the duration of the dry season. Sometimes it is sparse woodland, but often, as in the dry savannas of Africa and South America, the only vegetation consists of thorn bushes. Savanna country occurs in Africa, in southern Asia and in South America and its landscape changes according to the amount and distribution of rainfall. The wet savannas are characterised by over 1,200 mm of annual rainfall and a dry season lasting three to five months. They contain many species of trees and the grasses may stand several metres high. This type of savanna is found extensively in parts of Africa, especially in the west, in southern Asia and in South America. The somewhat drier savannas have average annual rainfall ranging from 500 to 1,100 mm and a dry season of five to eight months. These look more like parks, with well separated trees. This is the typical savanna of eastern and southern Africa, and is also found in South America. The true dry savannas or thorn bush savannas are dotted with species such as acacia and mimosa as well as trees adapted to drought, such as the baobab. Here the average annual rainfall is between 200 and 700 mm and the dry season lasts over eight months. This type of savanna is found in the Sudan, in Mali (the Sahel) and also in South America.

▲ Aardvark (*Orycteropus afer*)

▼ Troop of baboons (*Papio*)

▲ Honey badger or ratel (*Mellivora capensis*)

▼ African hunting dog (*Lycaon pictus*)

The African savannas

The mammal population of the African savannas is the best known wild animal community in the world. It includes enormous herds of antelopes, gazelles, giraffes, zebras, buffalos and rhinoceroses, together with their natural predators – lions, cheetahs, hunting dogs and scavengers such as jackals and hyenas. Basically there are three types of African savanna. In central and west Africa the average altitude is fairly low, around 300–400 metres. In east Africa it lies at the foot of the highest peaks on the continent and on either side of the great natural depression of the Rift Valley, at about 1,000 m, with a lower incidence of rainfall than in the west. Finally, there are the savannas of South Africa, also at an average altitude of 1,000 m and with even lesser amounts of rain than in the east and west.

▼ Cape jumping hare (*Pedetes capensis*)

Almost all the mammalian orders are represented in Africa. There are a number of insectivores such as hedgehogs, shrews and elephant or jumping shrews (*Elephantulus*). These are small, graceful creatures with an elongated, trunk-like muzzle and limbs that are well developed and adapted for jumping. Among insectivores, the aardvark (*Orycteropus afer*), is the only species belonging to the indigenous order Tubulidentata. Found all over Africa south of the Sahara, this strange nocturnal animal feeds mainly on ants and termites which it catches with its long, sticky tongue. It has powerful claws for destroying the termite mounds. It constructs remarkable burrows and these are often used by other mammals such as warthogs and hyenas. Termites and ants

▼ Black-backed jackal (*Canis mesomelas*)

also form the staple diet of the Pholidota, represented here by two ground species of the genus *Manis*. These are strange armoured anteaters common to Africa and southern Asia. This type of food is also eaten by the aardwolf (*Proteles cristatus*), a timid, nocturnal relation of the hyena.

There are a large number of carnivores roaming the African savanna. Mustelids include the ratel or honey badger (*Mellivora capensis*), the zorilla or striped polecat (*Ictonyx striatus*) and the striped weasel (*Poecilogale albinucha*). The Viverridae comprise about sixteen species of mongooses and civets. The Canidae are represented by the pale sand fox (*Vulpes pallida*); the bat-eared fox (*Otocyon megalotis*), a small nocturnal insectivore; the

▼ Spotted hyena (*Crocuta crocuta*)

African hunting dog (*Lycaon pictus*), an exceptionally fierce predator of all ungulates, hunting in packs; and three jackal species. These are the common jackal (*Canis aureus*), the side-striped jackal (*C. adustus*) and the black-backed jackal (*C. mesomelas*). Apart from the aardwolf there are three species of hyena, the brown hyena (*Hyaena brunnea*), the striped hyena (*H. hyaena*) and the spotted hyena (*Crocuta crocuta*). The Felidae include the serval (*F. serval*), the caracal (*F. caracal*) and the bush cat or caffer cat (*Felis libyca*), probably the ancestor of all our domestic cats. Large predators include the lion (*Panthera leo*), the leopard (*P. pardus*) and the cheetah (*Acinonyx jubatus*), fastest of all land mammals.

Among the smaller herbivores are several species of hare (*Lepus*), the scrub hare (*Poelagus marjorita*) and numerous rodents of all sizes, such as the ground squirrels of the genus *Xerus*, a few Cricetidae such as the Gambian pouched rat (*Cricetomys gambianus*), some porcupines, and the Cape jumping hare (*Pedetes capensis*) from southern Africa.

The majority of savanna herbivores are ungulates, perissodactyls and artiodactyls. There is also one proboscidian,

▼ African elephant (*Loxodonta africana*)

▼ Grevy's zebra (*Equus grevyi*)

▲ White rhinoceroses (*Ceratotherium simum*) ▶

the African or bush elephant (*Loxodonta africana*), largest land
mammal on earth. This elephant has massive ears and forward
curving tusks. It is well distributed in the eastern, central and
southern parts of the continent, but is no longer found in the
north. Another order found especially in the Oriental savannas is
the Hyracoidea. These hyraxes belong to three genera, *Procavia*,
Heterohyrax and *Dendrohyrax*, and also inhabit deserts and forests.

The savanna perissodactyls comprise members of the Equidae
and Rhinocerotidae. The Equidae include the numerous East
African zebras. The common or Burchell's zebra (*Equus
burchelli*) has a wide distribution and the species is divided into

several subspecies which include Grant's zebra, Selous's zebra and Chapman's zebra, ranging from Ethiopia down to South Africa. Grevy's zebra (*Equus grevyi*) is native to Somalia, Ethiopia and northern Kenya. It is more densely and delicately striped than the common zebra and has a white belly. It favours drier habitats and also lives in smaller herds. In the mountainous regions of South Africa there are still a few surviving mountain zebras (*Equus zebra*), but this original form is almost extinct.

A few Somali wild asses (*Asinus asinus Somalicus*) remain in the arid savannas, but the subspecies, the Nubian wild ass (*Asinus asinus nubicus*) is thought to be extinct. The African rhinoceroses are represented by two species, both with two horns, and, unlike the Asiatic rhinos, they still seem to be flourishing in fair numbers. The black rhinoceros (*Diceros bicornis*) is the smaller, its pointed upper lip being well suited for browsing on leaves and shrubs. There are estimated to be about 12,000 individuals inhabiting various parts of central and southern Africa. The white rhinoceros (*Ceratotherium simum*) is considerably larger. Weighing more than 3,500 kg, it is the second largest land mammal on earth. The white rhino has a flattened, trunk-like upper lip, adapted for grazing on grass. Today it is restricted to a few sections of central Africa and Natal, numbering only about 4,000 head.

▼ Waterbuck (*Kobus ellipsiprymnus*)

Thomson's gazelle ▶
(*Gazella thomsoni*)

Artiodactyls found in the driest savanna zones include the warthog (*Phacochaerus aethiopicus*). This boar, with its huge tusks, is capable of defending itself against predators. The Giraffidae are an endemic African family and the giraffe (*Giraffa camelopardis*) is a tall ruminant with an exceptionally long neck. It has two to five small horns covered with velvety skin, and its prehensile tongue is adapted for feeding on leaves and twigs. There are several subspecies, but the reticulated giraffe, from north-eastern Africa, is sometimes regarded as a separate species (*Giraffa reticulata*).

There are many savanna antelopes, including numerous species of the Tragelaphinae, a subfamily of the Bovidae. The elands (*Taurotragus*) are the largest antelopes, and the genus *Tragelaphus* contains various species, such as the kudus, the nyalas, the sitatunga or marshbuck, and the harnessed antelope or bushbuck. The subfamily Hippotraginae comprise a few species of large antelopes with long horns – the oryxes (*Oryx*), the roan and sable antelopes (*Hippotragus*), and the Reduncini. This last group contains several genera, the best known of which are the kobs (*Kobus*) and the reedbucks (*Redunca*).

The Alcelaphini subfamily includes many species of hartebeest (*Alcelaphus*), bastard hartebeest (*Damaliscus*), among them the topi, the blesbok and the bontebok, and the gnus or wildebeest

▼ Common eland (*Taurotragus oryx*)

▲ African or Cape buffalos (*Syncerus caffer*)

▲ Blackbuck or Indian antelope (*Antilope cervicapra*)

(*Connochaetes*). Members of the subfamily Antilopinae comprise all the gazelles. The most prominent among these are the gerenuk (*Lithocranius walleri*), the impala (*Aepyceros melampus*), the dibatag (*Ammadorcas clarkei*), Grant's gazelle (*Gazella granti*), Thomson's gazelle (*G. thomsoni*) and Soemmering's gazelle (*G. soemmeringi*). The Neotragini form a tribe of this subfamily and contain the smallest living Bovidae, the sunis (*Nesotragus*), the oribi (*Ourebia ourebi*), the steinbok (*Raphicerus campestris*) and the dik-diks of the genera *Rhynchotragus* and *Madoqua*.

The only species of cattle in Africa is the African or Cape buffalo (*Syncerus caffer*). The savanna form, found in the east and south, is a massive animal with a black coat and horns joined at the base.

Monkeys of the African savanna include some tree and some ground species. The tree monkeys are more abundant, but harder to observe. The galagos or bushbabies, are nocturnal primates living in groups. The commonest species is the Senegal galago or bushbaby (*Galago senegalensis*). The green monkey or vervet (*Cercopithecus aethiops*) lives almost exclusively in the savanna. The baboons (*Papio*) are all mainly terrestial creatures, the heavy-

▲ Jungle cat (*Felis chaus*)

bodied Cercopithecidae living in large troops with an elaborate social structure.

The Asiatic savannas
The Asiatic savannas are restricted to certain dry regions of Pakistan and central India. The few species of ungulates includes the Chinkara gazelle (*Gazella bennetti*) and the blackbuck or Indian antelope (*Antilope cervicapra*), with long spiralling horns, which lives in small herds. The blue bull or nilgai (*Boselaphus tragocamelus*) is now recognised as a primitive member of the Bovidae and a true antelope, related to the unusual four-horned antelope (*Tetracerus quadricornis*) which is the only wild representative of the Bovidae to carry four horns.

The savanna predators include the caracal, also found in Africa, and the Asiatic lion (*Panthera leo persica*), now reduced to only a few hundred animals in the Gir Forest reserve on the Kathiawar peninsula in north-west India. This is a dry, wooded region surrounded by subdesert steppe in which the ungulates are chiefly domestic cattle. The leopard has also been much reduced in numbers, though more frequent in forest zones, and the once familiar cheetah, now seems to have disappeared. There are two smaller members of the Felidae. The jungle cat (*Felis chaus*), found from Asia Minor across to the Indo-Chinese peninsula, as

▲ Nine-banded armadillo (*Dasypus novemcinctus*)

well as in Egypt, is a creature of forests and swamps in addition to savannas. The rusty-spotted cat (*F. rubiginosa*), inhabits southern India and Ceylon. Both species are light brown with irregular markings. They are larger than the European wild cat, preying mainly on birds such as pheasants or small mammals.

The South American savannas
There are various types of South American savanna which cover vast areas to the north and south of the tropical Amazonian rain forest or selva.

The mammals living here obviously occupy niches similar to those of the African savannas and there are many parallel species, although the large ungulates of the family Bovidae are absent in South America and replaced by a few Cervidae. They include the brockets (*Mazama*) and the marsh and pampas deer of the genus *Odocoileus*. Herbivores, fairly small in size, are represented by several rodents, including the paca (*Cuniculus paca*). Of the Suidae, the peccaries (*Tayassu*), although not exclusive to the savanna, are the counterparts of the African warthog. The small marsupials of the family Didelphidae correspond to the Old

▲ Paca (*Cuniculus paca*)

World insectivores, and the giant anteater and armadillos are very similar to the aardvark and pangolins. The giant anteater (*Myrmecophaga tridactyla*) is a characteristic inhabitant of the savanna, although sometimes straying into the forest or high plain. The tamandua or collared anteater (*Tamandua tetradactyla*) is sometimes seen in the savanna, but is essentially a forest animal. The armadillos, comprise species that roam both the savannas and steppes. The nine-banded armadillo (*Dasypus novemcinctus*) has a very wide distribution in America, being found as far north as Kansas, and seems equally at home either in

▲ Polar bear (*Thalarctos maritimus*)

arid regions or in rain forests. The giant armadillo (*Priodontes giganteus*) inhabits both rain forest and savanna.

In South America the carnivorous role of the leopard is assumed by the jaguar (*Panthera onca*), while the cheetah has its counterpart here in the maned wolf (*Chrysocyon brachyurus*). Both these carnivores are also found in the pampas. Other savanna carnivores, which also roam the high plains, the desert and the forests, are the grey fox (*Urocyon cinereoargentatus*), the bush-dog (*Speothos venaticus*), the grisons (*Grison*), and Felidae such as the ocelot (*Felis pardalis*) and the puma (*F. concolor*).

The tundra

The tundra is a belt of treeless land in the northern hemisphere, lying between the taiga and the Arctic ice-cap. The tundra covers several millions of square kilometres in Siberia, Russia and North America. There is a similar zone in the southern hemisphere, between latitudes 56° and 40°S, at the tip of Chile and Patagonia and embracing certain islands in Antarctica. This southern zone is less extensive than the Arctic tundra. From an ecological viewpoint, comparable zones, known as alpine tundras, are also found in mountain regions beyond the tree-line, but the fauna of these areas will be described in the section on mountains.

The basic physical factor regulating all forms of life in the tundra is not the quantity of water, but the low temperature. The

182

tundra is a wet, undulating plain, its soil being permanently frozen down to a depth of about one metre. Only the surface layer of earth thaws in the summer. Since the ice renders the soil impermeable, numerous small ponds and swamps form during the summer months. Other fundamental features of the climate in these parts are the extremely long winters, when the temperature may drop below −50°C, and the short summers, with temperatures averaging 10°–12°C. There is very little rainfall but the humidity level is high due to the lack of evaporation. There are also very strong winds, and the region is in total darkness from December until February.

Despite these harsh conditions animal and vegetable life quickly come to life during the brief summer season. The amount of food present at this time is sufficient for all the permanent mammal inhabitants of the tundra as well as the huge populations of migrating birds that come here to nest. The former include herbivores and predators which are continuously on the move in search of food both in winter and summer. No species of tundra mammal hibernates in the true sense. One reason for this is that

▼ Wolf (*Canis lupus*)

183

▲ Reindeer (*Rangifer tarandus*)

the summer is too short for an animal to accumulate the necessary stores of fat for a prolonged period of torpor. However these mammals have adaptations which enable them to survive. The fur is thicker, especially in winter, and particularly among the smaller creatures. Some mammals have a covering of hair on the soles of the feet which acts not only as insulation but also makes it easier to walk on snow and ice; and a subcutaneous layer of fat also keeps the animal warm. Modifications of hooves (as in reindeer) and claws (as in lemmings) also assist the process of walking over deep snow and ice. Finally, the summer coat may differ in colour from the winter coat. The Arctic fox, the weasel, the ermine, the Arctic hare and the collared lemming all turn white in winter. Reindeer and some populations of wolves also have paler coat colours in winter. The majority of small rodents (other than some species of lemmings and voles) retain their dark coats throughout the year. This is because they spend the winter in tunnels deep under the snow. The collared lemming (*Dicrostonyx torquatus*) has a white winter coat because it normally lives in shallow burrows close to the snow surface. Another common feature of tundra mammals is the shape of their body which is generally compact, with shortish limbs, tails and ears. The ears are also rounded and covered with thick fur.

▲ Musk-ox (*Ovibos moschatus*)

There are only two species of insectivore in the tundra, both red-toothed shrews (*Sorex arcticus* in Eurasia and *S. cinereus* in North America). Among the few rodents are the lemmings of the genera *Lemmus* and *Dicrostonyx* and the field-mice of the genus *Microtus*. Lagomorphs are represented by the snow or varying hare (*Lepus timidus*), although according to some experts the white hare of the tundra, which is also found in the taiga, constitutes a separate species (*Lepus arcticus*).

Tundra Cervidae are represented by the reindeer, known in America as the caribou, and the musk-ox. The Norwegian lemming (*Lemmus lemmus*) is the best known of the small rodents of the tundra. It is, however, a species that lives mainly in mountain regions. The more typical inhabitants of the tundra proper are the Siberian lemming (*Lemmus obensis*), which lives in colonies both in Eurasia and North America, and the previously mentioned collared lemming, with a more northerly distribution.

The ermine (*Mustela erminea*) and the weasel (*M. nivalis*) both roam the Eurasian tundra but one of the few carnivores really adapted to these regions, with specialised features that have already been listed, is the Arctic fox (*Alopex lagopus*). It excavates long, complex tunnels and preys chiefly on lemmings and voles. The largest tundra carnivore is the polar bear

185

(*Thalarctos maritimus*), which is found in both polar regions. It hunts seals as well as tundra reindeer, foxes and rodents. The reindeer (*Rangifer tarandus*) is the principal ungulate of the Arctic region, with a distribution similar to the polar bear. It visits the tundra during the summer and migrates for shelter to the taiga in winter. The musk-ox (*Ovibos moschatus*), though somewhat rare, still lives in Greenland and Alaska. This strange ruminant of the subfamily Caprinae looks like a true ox despite being more closely related to goats and sheep. Its horns are joined at the base to form a boss on the forehead, and it has long, shaggy hair.

▼ Rocky Mountain goat (*Oreamnos americanus*)

The mountains

Mountains can be considered as separate regions because their climates and therefore their habitats are different to those of the surrounding regions. The climates of mountain regions have a powerful influence on the vegetational pattern in adjacent areas and consequently on the composition of mammal populations. The distribution of plant life at successive levels of altitude to a certain extent corresponds to the pattern of vegetational growth at varying latitudes of the earth's surface. For example, a typical European mountain range has its foot forests surrounded by flat

▼ Chamois (*Rupicapra rupicapra*)

▲ Snow leopard or ounce (*Panthera uncia*)

areas of grassland, which are inhabited by grassland mammals such as wild boars, red and roe deer, and squirrels. At higher altitudes there is an imperceptible transition to coniferous forest and eventually, in the so-called alpine zone above the tree-line (uppermost limit of trees) there is a region of high prairie with typical mammal species. In Europe this zone is the haunt of the chamois (*Rupicapra rupicapra*), the ibex (*Capra ibex*), and the

▼ Chinchilla (*Chinchilla chinchilla*)

alpine or variable hare (*Lepus timidus*), already mentioned among the animals of the tundra. Rodents include the marmot (*Marmota marmota*) and the snow vole (*Nicrotus nivalis*). The only characteristic predator of this high zone is the ermine (*Mustela erminea*). However, it is not uncommon to find species from lower levels, such as the fox, the wolf and the brown bear, roaming at these heights.

The Bovidae of the sub-family Caprinae are also found in the mountain ranges of Europe and Asia. Thus the ibex (*Capra ibex*) is represented by several races in the Pyrenees, the Alps, the Caucasus, central Asia and Siberia. Similar, closely related races are to be found in the high mountains of east and north Africa. The European wild goat (*C. aegagrus*), probably the ancestor of domestic goats, was once very common, but is nowadays only

Ermine (*Mustela erminea*) ▼

◀ Varying hare (*Lepus timidus*), with summer coat (above) and winter coat (below)

189

found among the Greek islands and Asia Minor. The handsome markhor (*C. falconeri*) comes from central Asia. The mouflon (*Ovis musimon*) is found in the mountain ranges of the Mediterranean, notably Sardinia and Corsica. Asia is the home of the argali (*Ovis ammon*), which has a dozen or so subspecies including Marco Polo's sheep (*O. ammon poli*), from Pamir. In north Africa the Barbary sheep (*Ammotragus lervia*) is the only sheep native to the continent. Another mountain dweller, found in North America and Siberia, is the bighorn (*Ovis canadensis*).

▼ Ibex (*Capra ibex*)

Mention has been made of the chamois, with subspecies in Europe and Asia Minor. Related to these animals is the Rocky Mountain goat of North America (*Oreamnos americanus*). Other Bovidae living in the high mountain regions of Asia are the takin (*Budorcas taxicolor*), a strange animal from Bhutan and

southern China; the bharal (*Pseudois nayaur*) from the
Himalayas; the tahrs (*Hemitragus*); the goral (*Naenorhedus
goral*); and the serows (*Capricornis*). The serows are inhabitants
of southern China, south east Asia and Malaysia, with a single
species confined to Japan and Taiwan. The chiru (*Pantholops
hodgsoni*), an inhabitant of the high plateaus of Tibet and Ladakh
is related to the saiga of the steppes.

Other North American mountain animals include the mule
deer (*Odocoileus hemionus*), the black bear (*Ursus americanus*),

▼ Giant panda (*Ailuropoda melanoleuca*)

the mountain beaver (*Aplodontia rufa*) and the delightful raccoon (*Procyon lotor*). Rodent species of high mountain zones include various species of marmots, both in Asia and North America.

The mountain carnivores include Asiatic species such as the magnificent snow leopard or ounce (*Panthera uncia*) and the Himalayan black bear (*Selenarctos thibetanus*). The red or lesser panda (*Ailurus fulgens*) from Burma, China, Nepal and Sikkim, and the giant panda (*Ailuropoda melanoneuca*), live in the bamboo forests of southern China. Although mainly vegetarian, these pandas are occasionally carnivorous.

The fauna of the South American mountain ranges is more individual. Exclusive inhabitants of the Andes are the Camelidae of the genus *Lama*. Wild species are the guanaco (*Lama guanicoe*), probable ancestor of the domesticated llama (*L. glama*) and alpaca (*L. pacos*); and the rare vicuña (*L. vicugna*). Other typical residents of the Andean region are several Cervidae such as the guemals (*Hippocamelus*) and pudus (*Pudu*). There are numerous rodents, including the chinchilla (*Chinchilla chinchilla*), restricted to the highest and most inaccessible peaks.

The deserts

The deserts of the world extend for approximately 18 million square kilometres in areas where annual rainfall averages less than 250 mm. The principal deserts are in Asia (Arabia, Turkestan, Mongolia, Pakistan and China); Africa (the vast Sahara, and the Namib); the Americas (southern United States, Peru and Chile); and central Australia.

The predominant feature of all deserts is the scarcity of vegetation which never covers more than 50 per cent of the whole surface. Temperature fluctuates according to zones and there is a general distinction between cold and hot deserts. The deserts of central Asia have a typically continental climate with very low temperatures in winter and generally hot summers. Most of the mammals of these deserts are rodents, handy at digging and always building up food reserves, as is the case with the jumping mice of the genera *Rhombomys* and *Spermophilopsis*. The best known desert rodents, both in Asia and in other deserts, are the jerboas (*Dipus*, *Paradipus*, *Jaculus*) of the family Dipodidae, many of which can move extremely fast.

Some of these desert rodents are capable of existing entirely without water. Like many animals found in the desert, they are sandy in colour. Adaptations include long claws and long hair which help to avoid sinking into the deep sand, and also provide efficient insulation against heat. In the deserts of Kazakhstan, in central Asia, a species of rodent feeding exclusively on insects was discovered as recently as 1938. This is the unique Selevine's mouse (*Selevinia betpakdalaensis*).

There are several races of hare found in these deserts. Among the ungulates are the goitred or Persian gazelle (*Gazella subgutturosa*), which can race along at 80 km per hour. The saiga is sometimes seen in border areas. The Asiatic wild ass (*Equus hemionus*) is now represented by scattered populations, considered to be subspecies, such as the onager of Iran and

▼ Bactrian camel (*Camelus bactrianus*)

Afghanistan, the Mongolian kulan and the Tibetan kiang. The few carnivores include the sand cat (*Felis margarita*), the corsac or desert fox (*Vulpes corsac*), and a strange mustelid, the marbled polecat (*Vormela peregusna*). In one confined area of the Gobi Desert there still seems to be a wild population of the Bactrian camel (*Camelus bactrianus*), a species that was domesticated by man and became more widely distributed.

In the African deserts the most celebrated of many small rodents is the Egyptian jerboa (*Jaculus jaculus*). There are several small carnivores such as the pale sand fox (*Vulpes pallida*) and related species, the sand cat, and the fennec (*Fennecus zerda*), a small nocturnal fox with enormous ears. Among the Viverridae is a type of mongoose (*Suricata suricatta*), from South Africa. Desert ungulates include various species of oryx. The Arabian oryx (*O. leucoryx*), nowadays almost extinct, was once abundant in the Arabian deserts. The scimitar oryx (*O. dammah*), with its distinctive horns, is an inhabitant of the Sahara, and the gemsbok oryx (*O. gazella*) is found in South Africa. The addax (*Addax nasomaculatus*) is an antelope of the Sahara, as is the slender horned gazelle (*Gazella leptoceros*). Both species are now unfortunately very rare. The Dorcas gazelle (*G. dorcas*) appears to have a more hopeful future and is fairly abundant both in the

▼ Oryx (*Oryx gazella*)

▲ Egyptian jerboa (*Jaculus jaculus*)
▼ Onager (*Equus hemionus onager*)

▲ European mole (*Talpa europaea*)

Sahara and the Sachel. Finally, in the mountain zones of African deserts the rock hyrax (*Procavia capensis*) lives in large groups.

The deserts of North America comprise the searingly hot Death Valley region in the centre of the Rocky Mountains and the vast plains of Arizona and Colorado. As far as mammals are concerned, there are no ungulates but there are numerous rodents, including kangaroo-rats very similar to the Old World jerboas. Others are the spiny pocket mice (*Perognathus*) and various ground squirrels (*Citellus* and *Ammospermophilus*). Lagomorphs include the white-sided jack rabbit (*Lepus alleni*) and the Audubon rabbit (*Sylvilagus auduboni*). Carnivores include the kit fox (*Vulpes velox*), which is sometimes joined by the coyote and the puma. There is also an insectivore, Crawford's desert shrew (*Notiosorex crawfordi*).

Underground mammals
Certain mammals live underground, in rock clefts or heaps of large stones, among tree roots or in burrows. These burrows can sometimes be very elaborate. Some species dig their own burrows, while others make use of existing ones. Most of these mammals, however, regularly leave their shelters and go outside in search of food. Compared with mammals that live outdoors they have relatively few physical adaptations, apart from large nails transformed into claws for digging. However, there are a few

▲ Naked mole rat (*Heterocephalus glaber*)　　　▲ European mole rat (*Spalax microphthalmus*)

species that live permanently underground, and these animals have specialised adaptations for their subterranean life.

Animals that live partially or totally underground are able to move around in one of two ways. Some have long, slender bodies, pointed at the tip, and are able to thread their way through particles of earth. This type includes earthworms, certain subterranean insects, the worm lizards (Amphisbaenidae) and the blind snakes (Typhlopidae). Others dig a tunnel or gallery with natural tools such as teeth or claws. The body shape of these creatures is usually squat and rounded and movement is kept to the minimum. This form of locomotion is employed by the mole-cricket, various types of underground insects and by the mole.

All mammals that live permanently underground adopt the second type of locomotion. Their body is short, stocky and rounded, the tail is little more than a stump, and the eyes are either very small or wholly absent, as are the ear auricles. The coat consists of velvety hairs that do not point in any particular direction, enabling both forward and backward movement through narrow tunnels. The digging tools are the forelimbs, which are short and shaped like shovels, with powerful muscles and strong claws – as in the case of marsupials and insectivores. The rodents have huge, sprouting incisors.

The network of galleries excavated by these creatures form true territories around the burrow or nest. The soil removed during digging is piled up outside, forming mounds. These are common sights in fields, where they betray the presence of moles; on steppes, where they are the work of mole rats; and on savannas where African mole rats live.

In their branching subterranean galleries insectivores feed on earthworms and insect larvae while rodents eat roots, bulbs and tubers. In the desert flatlands of central Australia the marsupial mole (*Notoryctes typhlops*) is a little known animal with golden fur, two huge claws on the forefeet, and a very short tail. It is outwardly very similar to the South African golden moles (Chrysochloridae). The latter are insectivores, differing from true moles by virtue of their colour, the absence of a tail and the presence of only two or three large nails on the forefeet. Another animal similar to the mole is the rice tenrec (*Oryzorictes talpoides*) of Madagascar, which digs galleries under rice fields.

The most numerous and widely distributed of the insectivores are the Talpidae. This family embraces the true moles belonging to the genera *Talpa, Scalopus, Condylura* and others. All these species have five strong claws on the forefeet and in more specialised forms, such as the European mole (*Talpa europaea*), there is an additional half-moon-shaped bone in the hand which increases its width. The muzzle is pointed, and strengthened by prolonged nose cartilages. In the star-nosed mole (*Condylura cristata*) the muzzle bears a sensory structure of star-shaped tubercles. The tail is rather short, with tactile hairs.

Among the rodents there are a few species of the family Geomyidae (the American pocket-gophers) and of the Ctenomyidae (tuco-tucos of South America). Both species use their foreclaws and large incisors for digging. Some subterranean mice also make good use of their claws, while the lemmings of the genus *Ellobius* and all the more specialised rodents only use their teeth. These incisors are very well developed and always appear on the outside, even when the lips are closed. This prevents the mouth from becoming full of earth. The most typical rodents are representatives of African family Bathyergidae, the mole rats, with tiny eyes, short legs and skimpy tail. Comprising several genera, the family has a particularly strange member in the naked mole rat (*Heterocephalus glaber*) of East Africa, which lives permanently underground in deep tunnels. The entrances of these tunnels are marked by high mounds of removed soil. It is an extremely ugly animal, almost completely without hair, pale pink in colour and a wrinkled skin.

The mole rats of the family Spalacidae are represented by a few species living in the Mediterranean region and other parts of Europe and Asia. They are the most specialised of all the

Flying foxes (*Pteropus*) roosting in a cave ▶

▼ Flying fox (*Pteropus*)

◀ Flying squirrel (*Glaucomys*)

subterranean rodents for they are wholly blind, being completely without eyes. The tail and ear auricles are also practically non-existent. These animals excavate galleries with their huge incisors and fling the loose earth backwards with their feet. Like moles, they live underground all the time. Their tunnels are very complex structures and can reach a depth of three metres.

▼ Grey gibbon (*Hylobates moloch*)

Among other underground inhabitants are animals which make use of caves and hollows. Although there are no true cave-dwelling mammals, many species make use of artificial or natural shelters either spending part of the day resting there or resorting to such refuges in periods of torpor. All of them, however, venture outside for food. In parts of southern Europe caves provide shelter for various rodents, including rats, and sometimes for carnivores such as the marten and the fox. In North America caves provide temporary refuge for certain meadow voles (*Microtus*) and wood rats (*Neotoma*).

In the air

Certain climbing animals that live in trees and have to cross from branch to branch are able to leap through the air and are equipped with fairly specialised structures to assist them in a form of gliding flight. Many arboreal creatures manage to move about with great agility in the trees with the aid of special adaptations. These include a prehensile tail, opposable fingers, hands transformed into hooks, or fingers with suckers. None of these are capable of gliding but some marsupials, rodents and lemurs hurl themselves from one branch to another with outstretched limbs, with the aid of a cutaneous membrane which expands into a kind of wing. The only mammals which can truly fly are the bats.

Some marsupials of the family Phalangeridae, from Australia and adjacent islands, glide with the aid of this membrane, or potagium, which extends on either side of the body, linking the fore and hind limbs. Additional assistance is given by a fringe of long hairs on the sides of the tail, as in the gliders of the genus *Acrobates* or by an especially thick tail, as in those of the genus *Petaurus*. Certain rodents such as the flying squirrels of the genera *Petaurista, Aeromys, Pteromys*, have structures very similar to the marsupials (membrane and flat, bushy tail). Even the common squirrel has a rudimentary form of broad, bushy tail. Cutaneous folds on both sides of the body enable the flying squirrels to glide gracefully. Such animals are active at dusk and during the night, leaping from branch to branch and sometimes even using air currents to vary their flight trajectory. In this way they can glide for several hundred metres. The Cynocephalae – the so-called flying lemurs, although they are not true lemurs and cannot really fly – possess a very large membrane which extends from head to tail and outwards to the toes and fingers.

A structure similar to the membrane of these gliders, but extending between the elongated fingers is seen in those mammals that have attained genuine flight capacity, the Chiroptera or bats. Here the patagium has been transformed into a proper wing that

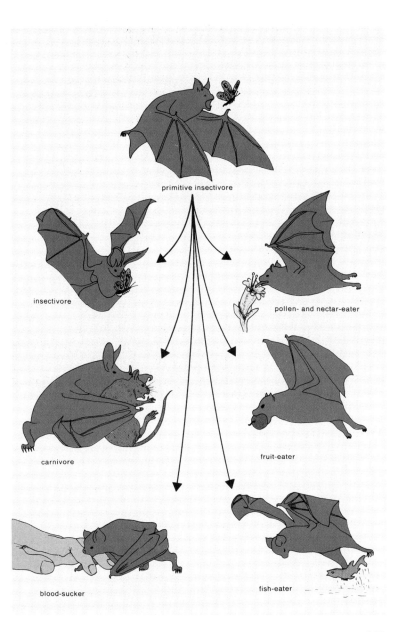

primitive insectivore

insectivore

pollen- and nectar-eater

carnivore

fruit-eater

blood-sucker

fish-eater

▲ Various species of bats have different feeding habits

▲ Vampire or blood-sucking bat (*Desmodus rotundus*)

can be beaten to and fro to produce a real flying action. Naturally, not all bats are anatomically alike. In some species the tail may be free and also very long, while in others the tail is partially or wholly contained in the patagium. The wing form of the various bats also shows considerable flexibility. Species that flit in and out of trees have broad, rounded wings rather like those of sparrowhawks. Others have long, narrow wings like those of the true falcons. As a result of their being specialised fliers, the skeletal structure of bats shows certain modifications. This is evident in the enormously developed fingers, the toes and the hind feet which are modified for clinging, the presence of a large clavicle and of a keeled sternum to accommodate the strong breast muscles activating the wings.

▲ Flying foxes (*Pteropus*)

The Chiroptera comprise many species diffused throughout the world, including New Zealand. They are extremely specialised creatures with individual food preferences. Most forms are insectivores and these nowadays make up the majority of species in temperate zones. Some bats prey on small mammals and fishes, while the vampire bats of central and South America suck the blood of large mammals, using their sharp incisors to inflict painless wounds while the animals are asleep. Finally there are numerous species that eat fruit, and some that suck nectar and pollen from flowers by means of an elongated snout and tongue, hovering in the manner of humming birds.

Most of the bats that feed on insects are active after sunset. They spend the day hidden in natural cavities, sometimes in

▲ Walrus (*Odobenus rosmarus*)

caves. Although certain species migrate to warmer climates during winter to prey on insects, the majority spend the winter months sleeping. Cave species may be solitary but are frequently gregarious, forming enormous, dense populations. The bats find their food outside but deposit quantities of guano inside the cave. In tropical regions piles of these droppings may cover the cave floor to a depth of several millimetres. Guano has a commercial use as a fertiliser and it represents the major energy link between the gloomy world of the cave, almost devoid of autotrophic plants, and the outside world.

Bats move around in the darkness by using a system of echolocation. They transmit high-frequency sound waves which bounce off objects and are received by a highly developed and specialised hearing system. By this means they can distinguish the nature of an obstacle and calculate its distance with extraordinary precision. This facility enables them not only to capture insects in total darkness but also to fly freely around the cave, avoiding collisions with stalactites and other natural hazards. The sound waves emitted by bats are of a very high frequency, ranging from 30,000 to 100,000 cycles per second. This is far beyond human hearing capacity.

It would seem that each species has its individual system of

▲ Sea elephant (*Mirounga*)

ultrasonics in respect both of length and frequency of pulses. The latter may also vary, increasing when the animal approaches close to an obstacle, decreasing as it gets farther away, so as to avoid any overlapping of sounds. The bats of the family Vespertilionidae emit pulses from the mouth, whereas the Rhinolophidae or horseshoe bats produce them through the nose which serves as a kind of funnel concentrating the sound into a narrow beam that can be directed by head movement. The ultrasonic receptor system consists of the ears. The ear structure varies in complexity according to species and often exhibits unusually large auditory lobes.

In the water

Many mammals are dependent upon a watery environment. They may live, wholly or partially, in the ocean, in the estuaries, salt marshes, lagoons and mangrove swamps, or in rivers, streams, lakes and ponds. Certain mammals live permanently in water and are anatomically and physiologically specialised for this type of habitat. Some live in water but sometimes emerge onto dry land; and yet others lead virtually an amphibious existence.

The basic features of aquatic life are the ability to swim and the capacity to avoid loss of heat in the surrounding watery

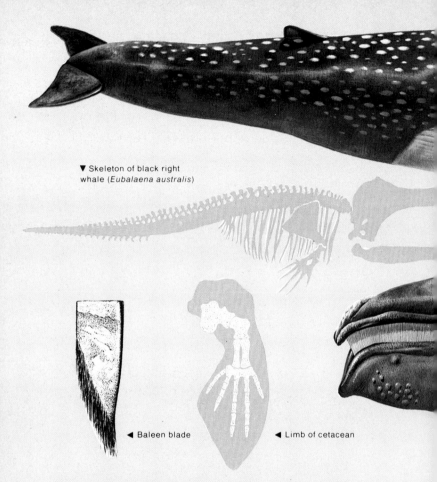

▼ Skeleton of black right whale (*Eubalaena australis*)

◄ Baleen blade

◄ Limb of cetacean

environment. Almost all mammals are able to swim. Some water animals are able to float more easily due to a thick layer of body fat which lowers their specific weight. This fatty layer is also important for heat insulation and in many of the more specialised animals, notably seals, walruses, sea elephants and whales it is extremely thick. However, these species have very little or almost no hair. The thickness of this adipose layer varies considerably according to an animal's way of life. In sea elephants and walruses it may be about 15 cm but in whales it is two or three times as thick. Naturally, the slower-moving whales, such as the baleen species, have a thicker layer of fat than the faster forms,

▲ Blue whale (*Balaenoptera musculus*)

▼ Black right whale (*Eubalaena australis*)

such as the rorquals, in whom it seldom attains 30 cm. The reason for this is that an insulating layer that is too thick might lead to overheating. Fat as a heat insulator is also a feature of the sea cows of large amphibious mammals and of semi-aquatic animals such as the hippopotamus.

Many aquatic mammals and amphibians have a coat of downy, soft, dense hair underlying the longer guard hairs. Examples of this are seen in the beaver, the otter, the polar bear and the fur seal. When such creatures are in the water only the guard hairs get wet and these protect the underlying downy layer which remains completely dry. When the animal comes out of the

▼ Sperm whale (*Physeter catodon*)

water a few vigorous shakes will rid the coat of moisture. Other mammals, including the seals and sealions, have a coat that gets thoroughly wet and it is these that need an insulating layer of fatty tissue. Species such as the platypus and various water shrews, have a soft coat of fur which retains air bubbles. However after a certain period of immersion in water the coat becomes completely drenched and the animals have to dry themselves in their own burrows along the river bank.

The majority of more specialised aquatic mammals live in the world's oceans. First and foremost come the numerous families, genera and species making up the order Cetacea, commonly known as whales. The body is fish-like, the forelimbs have been transformed into fins, and the hind limbs have disappeared. The expansion of the skin surface on either side of the tail has formed a tail fin and often a similar process on the back has shaped a dorsal fin. Other less obvious characteristics are internal adaptations to facilitate diving to great depths, a sonar-type orientation system and a specialised form of diet. Such is their adaptation to aquatic

▲ Killer whale or grampus (*Orcinus orca*)

▲ Narwhal (*Monodon monoceros*)

life that they are incapable of surviving on dry land where the weight of the body, not counterbalanced by hydrostatic pressure, would cause irreparable damage to the lungs and internal organs.

The present-day cetaceans are divided into two vastly differing suborders: the Mysticeti, comprising the baleen or whalebone whales, and the more abundant Odontoceti, or toothed whales, including sperm whales, dolphins, etc. The baleen blades of the Balaenidae and Balaenopteridae are specialised triangular, horny plates arranged on either side of the palate and inserted in the jaws. The inner side of each blade contains a fringe of bristles. These strain the sea water, retaining the plankton, comprising crustaceans, fish eggs and molluscs, which constitutes the baleen whale's food.

Many of the baleen whales have an enormous range of distribution, while others inhabit more limited ocean zones. Nearly all of the baleen whales have become very rare, with certain species on the verge of extinction as a result of excessive whaling. The pygmy right whale (*Caperea marginata*) is native to

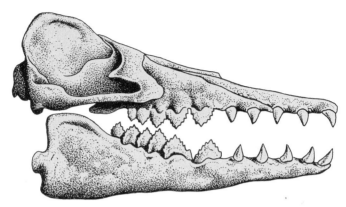

▲ Fossil skull of extinct cetacean suborder Archaeoceti

the seas around Australia. The Californian grey whale
(*Eschrichtius gibbosus*) is the only species of the family
Eschrichtidae (Rachianectidae). This inhabitant of the Arctic and
northern Pacific was thought to be almost extinct but now seems
to be making a comeback, having received a measure of
protection. The rorquals and related species of the family
Balaenopteridae have become extremely rare, though found in all
oceans. Among them is the blue whale (*Balaenoptera musculus*),
the largest mammal ever to have existed. Another is the strange-
looking humpback whale (*Megaptera novaeangliae*), with
exceptionally long pectoral fins. The right whales, represented by
three species in the Arctic and Antarctic, are now almost extinct.
They are slower and stockier than the rorquals, and have much
longer baleen blades.

▼ Skull of sperm whale
(*Physeter catodon*)

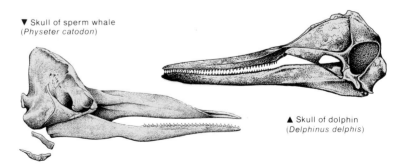

▲ Skull of dolphin
(*Delphinus delphis*)

▲ Harp or Greenland seal (*Pagophilus groenlandicus*)

The Odontoceti are found in much greater numbers than the Mysticeti. Their teeth are all alike, and sometimes very numerous, as in the La Plata dolphin (*Stenodelphis blainvillei*) which has about 250. In other species the teeth are far fewer or, in certain forms, completely absent. The Odontoceti include some very curious-looking genera and species, including whales with an asymmetrical head (possibly an advantage for sonar-type orientation) and others with unusual teeth, like the male narwhal (*Monodon monoceros*), from the Arctic which has one left tusk.

Closely related to the narwhal is the beluga or white whale (*Delphinapterus leucas*), also living in Arctic seas in large packs. The family Ziphiidae contains a number of beaked or bottle-nosed whales, some of which (*Hyperoodon* and *Berardius*) are fairly abundant and fished commercially. Others, such as Cuvier's whale (*Ziphius cavirostris*) and, above all, the whales of the genus *Mesoplodon* are extremely rare or only known from specimens stranded on beaches. In the *Mesoplodon* species there are only two teeth in the lower jaw. In one case they resemble the tusks of wild boars, and prevent the mouth from completely closing. Oliver's beaked whale (*Tasmacetus shepherdi*) known only from two skeletons, was discovered in 1937 on the coasts of New Zealand.

The Physeteridae contain two species. The sperm whale or cachalot (*Physeter catodon*) is the largest of the toothed whales with an enormous barrel-shaped head. Again the structure of the head is probably vital for sonar location under water at great depths. The sperm whale is found in all seas, particularly in hot

▲ Manatee (*Trichecus*)

and temperate latitudes. The pygmy sperm whale (*Kogia breviceps*), although present in all oceans, is much less familiar.

The other families of toothed whales, often linked in the single family Delphinidae, contain many species of different sizes, found all over the world. It includes the grampus or killer whale (*Orcinus orca*), a voracious predator roaming every ocean, which grows up to 9 m long. The dolphins, porpoises and pilot whales which also belong to this family include a number of abundant and well-known species, many of them gregarious. These include the common dolphin (*Delphinus delphis*), the bottle-nosed dolphin (*Tursiops truncatus*), the pilot whale (*Globicephala melaena*) and the common porpoise (*Phocaena phocaena*). Certain species of the genera *Stenella*, *Lagenorhynchus* and *Cephalorhynchus*, have occasionally turned up by accident in fishing nets but are extremely rare. The Bornean dolphin (*Lagenodelphis hosei*) was only discovered in 1956. Its description is based on a single skeleton found beached in 1895, and it has never been seen alive.

Other completely aquatic mammals, specialised for this type of life, are the Sireniia (sea cows), an order comprising a few exclusively vegetarian species which look much like the wholly carnivorous cetaceans. Similar features are the fish-like shape of the body, the forelimbs transformed into fins, the absence of hind limbs and the horizontally enlarged tail. The dentition is, however, quite different, with continuously growing incisors and molars. These animals are placid and lazy. There are four living species belonging to two families. Another species, Steller's sea cow

214

Seal among ice-floes ▶

(*Hydrodamalis stelleri*), was exterminated towards the end of the 18th century. It had been discovered only in 1741 in the Bering Sea. The three species of manatee, of the genus *Trichecus*, are, however, found along the tropical African and American coasts of the Atlantic and in the rivers flowing into that ocean. The dugong (*Dugong dugon*) inhabits the shores of the Indian and Pacific Oceans from East Africa to Australia and the Philippines. The difference in the dentition (manatees having the strongest teeth) is probably related to the fact that these animals feed on the more fibrous aquatic plants growing on land, whereas the dugong's diet consists mainly of weed-like marine algae.

The other marine mammals are the seals and their allies, the carnivorous Pinnipedia. Each family has its distinct manner of swimming. The true seals get their main thrust from the hind flippers which act like the tail of fishes and whales. The eared seals and walruses are propelled chiefly by the fore flippers. These mammals are found in seas all over the world, especially in the Arctic and Antarctic, but they are almost never seen in the Indian Ocean. The family Otariidae contains several species such as the sea-lions, which live off the shores of the American Pacific, and the fur seals or sea-bears, mainly found in the North Pacific. The Odobenidae consists of one species only, the walrus (*Odobenus rosmarus*). It is similar to the fur seals in its limb structure and method of swimming, but it has no ear auricles. The male walrus grows to a huge size and also boasts enormous tusk-like upper canine teeth which help to spear molluscs and crustaceans on the ocean bed. The species is found only in Arctic seas close to the coasts and around the polar ice-cap.

The true seals of the family Phocidae are the most perfectly adapted of the pinnipeds to marine life and comprise eighteen species, most of them in the northern hemisphere. Some, like the bearded seal (*Erignathus barbatus*), the harp seal (*Pagophilus groenlandicus*) and the ringed seal (*Pusa hispida*) are found in virtually all northern seas, whereas the grey seal (*Halichoerus grypus*) is confined to the North Atlantic. The common seal (*Phoca vitukina*) is also found in the North Pacific. A species typical of the southern hemisphere is the crab-eater seal (*Lobodon carcinophagus*) of Antarctica, which feeds on the same kind of shrimp-like plankton (krill) as the baleen whales. It strains its food by means of special structure of its molars. The monk seals inhabit tropical and subtropical seas. One species (*Monachus monachus*) is the only pinniped found in the Mediterranean, although the population is small, scattered and threatened with extinction. Apart from those already mentioned, other Phocidae found in Arctic seas include the hooded seal (*Cystophora cristata*)

▲ Common otter (*Lutra lutra*)

and the ribbon or banded seal (*Histriophoca fasciata*), so named because of the yellow bands on its flanks. The walrus, the narwhal, the beluga and some other whales are also native to these waters. The polar bear is often seen hunting and swimming in the ocean, sometimes a great distance from the coast.

The fur seal is an inhabitant of the North Pacific, around the Bering Sea. This region is also the home of one of the strangest of sea mammals, the sea otter (*Enhydra lutis*). A large mustelid, about 1.5 m long, with a short tail, the sea otter weighs up to 40 kg. Its thick, impermeable coat is extremely beautiful and highly valued. In the 19th century this species, which lived along the coasts from California to Alaska and from Kamchatka to the Curile Islands, was all but exterminated by fur-sealers. Today efficient protective measures have resulted in an increase in numbers and this handsome creature can again be seen in coastal waters. It lives almost permanently in the sea, and even sleeps in the water. It feeds on sea urchins, crustaceans, molluscs and various fishes.

217

Other types of otter live in fresh water. They are elegant creatures of various sizes. Most otters have short legs and webbed toes, sometimes without claws, and a powerful, elongated tail that is invaluable for swimming. They also have thick fur and small ears that can be closed under water. These delightful, sociable mammals feed on aquatic animals, fishes, amphibians and reptiles. They hunt either alone or in pairs. The European or common otter (*Lutra lutra*) is not as abundant as it used to be, having disappeared from many lakes and rivers partly as a result of hunting but even more because of water pollution which has had disastrous effects on the fish population. The otter has often been hunted on the grounds that it eliminates the fish population in certain rivers. In fact, it is highly valuable in helping to maintain the natural equilibrium of river life by getting rid of sick and weak individuals. Inland waters are the habitats of many specialised aquatic mammals. Lake Baikal is the home of an endemic species, the Baikal seal (*Pusa sibirica*), while the Caspian seal (*P. caspica*) lives in the Caspian Sea.

Some of the sirenians living in tropical zones often ascend rivers, appearing many miles from the coasts. Various cetaceans, especially the dolphins, frequently make their way up river estuaries. A few years ago a white whale was seen in the Rhine, and another specimen was caught in the lower St Lawrence.

Dolphins of the family Platanistidae live in fresh water and are to be found in the great rivers of Asia and South America. There are four species of four different genera. Growing to three metres long, they have a fairly broad head, a long, narrow snout and in some cases, numerous teeth. Their general appearance is rather odd: the dorsal fin is a low, keel-like projection; there is only a hint of a neck and the eyes are very small. The Gangetic dolphin (*Platanista gangetica*) is completely blind. This strange creature lives in the rivers of northern India (Indus, Ganges, Brahmaputra and their tributaries), either alone or in small groups. It hunts in the muddy river bed for fishes and crustaceans with its extremely sensitive, long snout and orients itself entirely by echo-location. A South American species, living in the river basins of the Orinoco, the Amazon and their tributaries, is the inia (*Inia geoffrensis*). This species has the characteristic projection and tactile hairs above the snout, but in all other respects lives like the rest of the freshwater dolphins. The remaining species are the Chinese river dolphin (*Lipotes vexillifer*), found only in one Chinese lake where it seems to be fairly abundant and which was described only in 1918, and the La Plata dolphin (*Stenodelphis blainvillei*) of the Rio de la Plata. The latter has a long snout, with up to 250 teeth, which is more than any other living mammal. In both these

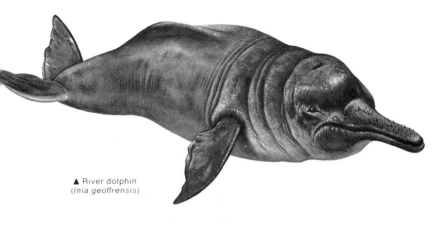

▲ River dolphin
(*Inia geoffrensis*)

species the dorsal crest is raised and more like a real fin. However, the La Plata dolphin does not live permanently in fresh water, for in winter it migrates along the coasts of southern Brazil and Argentina.

Finally, there are a number of less specialised aquatic animals, comprising rodents and insectivores. Among the former are the water voles (*Arvicola*), with fringes of hair on their hind feet and several species of water rats (*Hydromys*) from Australia. The

▼ Gangetic dolphin (*Platanista gangetica*)

▲ Coypu (*Myocastor coypu*)

round-tailed muskrat (*Neofiber alleni*) of Florida has incomplete webbing on the hind toes, and the muskrat or musquash (*Ondatra zibethica*) has a tail that is flattened on the sides, as well as partially webbed fore and hind feet with fringes of hair. The beaver (*Castor fiber* in Europe and *C. canadensis* in North America) has a flattened tail and completely webbed hind toes, as does the coypu (*Myocastor coypu*) and the fish-eating rats, *(Ichthyomys)* with fringes of hair on the sides of the feet and beneath the tail. Partial webbing is also present in species that are only semi-aquatic, as, for example, in the largest of the rodents, the capybara (*Hydrochoerus hydrochaeris*).

The insectivores include species with hair on the feet and under the tail, such as the water shrews of the genus *Neomys*, found in Europe, and those of the genus *Nectogale* in Tibet. Other species, in addition to hair, also have webbed toes. These include the desmans (*Desmana moscatha* from southern Russia and *Galemys pyrenaicus* from the Pyrenees). These animals have a tail flattened at the tip, in the same manner as an aquatic species of tenrec (*Limnogade*) from Madagascar. The most specialised of the water insectivores are the otter shrews (*Potamogale* and *Micropotamogale*) found in African streams and rivers. Their feet are not webbed but the powerfully developed tail is the main organ used for swimming.

Naturally there are other mammals belonging to various orders

▲ European beaver (*Castor fiber*)

which are in some manner adapted to life in the water, such as the hippopotamuses, a number of herbivores with amphibious habits, like the Indian rhinoceros, and a few antelopes and deer. Carnivores with aquatic preferences, apart from the otters, include the marsh or water mongoose (*Atilax paludinosus*), which finds its prey in streams and swamps, and the raccoon (*Procyon lotor*), from North America, which feeds on frogs and fishes.

Some of the rodents mentioned above supplement their plant diet with fishes, crustaceans and molluscs. The American fish-eating rats of the genus *Ichthyomys* feed almost wholly on fishes, and in the Mediterranean region the water vole consumes large bivalves of the genera *Unio* and *Anodonta*, cracking open the shells on rocks and stones.

Specialised aquatic forms are also found among the more primitive mammals, including the platypus and the marsupial known as the yapok or water opossum (*Chironectes minimus*). The yapok, from South America, has webbed feet but free toes. When under water it is able to close its pouch.

Mammals and Man

The previous chapters of this book have outlined the main structural and behavioural characteristics of this class of mammals. Before concluding with a chapter on mammal classification, it is worth devoting some space to the relatively long-standing relationship between man and other mammals.

In biological terms man is classified as an anthropoid ape of a particularly advanced type. What distinguishes him from other mammals is the complexity of his brain which gives him the power to reason, to use tools and to turn external forces to his advantage. The complex nature of his brain is one of the reasons why man has evolved so successfully and why he has been able to adapt to virtually every available habitat on the earth's surface. Man is not a mammal specialised for any particular type of existence. His ability to adapt to his surrounding environment has depended more on his ability to think and reason than on anatomical modifications. Within a relatively short period of time this so-called 'naked-ape' has populated and dominated the earth, changing both his way of life and his environment whenever and wherever necessary.

Although man is now earth's dominating mammal he has always maintained important relationships with other mammals. This is because, despite the extraordinary technological progress man has made over the centuries he has always depended on certain mammals for his very survival. As a primitive hunter he

merely stalked and ate the mammals he managed to kill. However, as he began to live in permanent dwellings, he also domesticated and bred animals. These animals not only ensured a continuous food supply but also assisted him in raising crops. The gradual process of domesticating mammals was not necessarily solely stimulated by the urge to make immediate, practical use of the animals concerned. It is interesting to note, for example, that one of the first, if not the very first, of the domesticated species was the dog. These early hunting companions probably derived from populations of wolves that prowled about human settlements hunting for food among the refuse. Other domesticated mammals, utilised by man for their meat, milk and skins, were invariably gregarious animals such as sheep, goats, reindeer and the like.

Domestic animals can be said to belong to various categories. Some, like the dog, probably came into contact with man by accident, more as companions than for any economic purpose. Secondly, there are the small carnivores, such as cats, ferrets and mongooses which naturally hunt animals that are in some way harmful to man, notably rats, mice and snakes. Finally there are all the mammals that provide man with meat, milk, hides, fur and so on. There are a large variety of mammals in this last category including cows, zebus, buffaloes, gayals, yaks, goats, sheep, pigs, reindeer and the various members of the camel family such as llamas, alpacas, guanacos and vicuñas.

▲ European bison (*Bison bonasus*

◀ Przewalski's horse
(*Equus przewalskii*)

◀ Preceding pages: herd of llamas and alpacas in South America

Many of these animals are also used for transport and as beasts of burden and perhaps some of the most important in this respect are the various races of horses. Today most horses are descended from the wild horses of central Asia. Asses are similarly derived from wild stock, which is now almost extinct. Other domestic animals no longer found in their wild state include the camel, the dromedary, the llama, the alpaca and the ox.

Certain animals tamed and used by man are captured from wild populations and are not bred regularly. The Indian elephant, for instance, is usually caught and trained only when it has reached adulthood. Other species such as the reindeer are only semi-domesticated. In this instance the relationship is between animals and people who are both nomadic and migratory. A similar situation in which man was dependent upon a single mammal species, but without any serious attempt at taming it, was that of the American bison previously mentioned in the section on migration. The European bison, on the other hand, never seems to have had as much importance to the people with whom it lived, and its virtual extinction was due to excessive hunting.

In spite of the fact that man relies heavily on certain species of mammal for his very existence, it is unfortunately often difficult for him to live harmoniously with many wild species. When wild animals raid or destroy man's crops, he often responds by killing the offenders. Sometimes only a few individuals are destroyed but in the past whole species have been exterminated. The list of mammals that have either vanished or been decimated by human action during the past couple of centuries is unfortunately, a long one. It includes cetaceans, from huge whales to small dolphins, and sirenians such as Steller's sea cow, which disappeared almost before it became known to science. Ungulates have also been among the principal victims. The blaubok (*Hippotragus leucophaeus*) and the quagga (*Equus quagga*) both used to roam the plains of southern Africa in huge herds, and both are now extinct as a direct result of Afrikaaner settlement. Similarly there are only small herds of Arabian oryx (*Oryx leucoryx*) and Arabian gazelle (*Gazella gazella arabica*), which are shot for sport, and live a precarious existence. Various species of African and Asiatic sheep and goats have been exterminated for the sake of their horns, displayed as trophies. Similarly the rhinoceroses of the islands of the Far East have been reduced to tiny groups.

The carnivores have also suffered from being hunted for their pelts. Victims include the mighty tigers, the South American otters, the small mustelids and the enormous bears. The brown bear, although still fairly abundant in the Carpathians and in Russia, is nowadays restricted to three or four individuals in the

Alps. There are also a few more, barely surviving, in the
Apennines and the Pyrenees. The range of the big cats has been
drastically reduced as a result of the hunting and poaching
activities of man. Only a few centuries ago the lion, for example,
enjoyed a wide distribution in the Mediterranean regions, in Asia
Minor and in Persia, but today is limited to the savannas of east
Africa and a small protected population in India. In Australia,
New Zealand, and Tasmania, man has imported new species
which have sometimes caused irrevocable imbalances to the
ecosystem. In some cases, man's interference with the natural
course of events has resulted in the total extinction of a race. A

▼ Domesticated Indian elephants enjoying a bath

prime example of this latter phenomenon is the extermination of a unique race of aborigines in Tasmania. These people lived in total isolation since ancient times and were only discovered by western man in the late 18th century. Within the next century this primitive race was completely wiped out. This was partly the result of the white men shooting the 'savages', and partly due to the fact that the aborigines had no resistance to white man's germs, notably the common cold.

Another important aspect of the relationship between man and other mammals is that some species are responsible for carrying various diseases which are harmful, or even fatal to man. A

notable example of this is where rabies is transmitted to humans via some species of bats, as well as dogs, cats and foxes. A wide range of tropical diseases, including yellow fever and lasser fever, are harboured by various monkeys.

It is possible to argue that the extinction or local disappearance of an animal species is a fact of life, inextricably linked with evolutionary development. However, with the technological

▲ Whales laid out for processing at a shore station

means at his disposal today, man has the ability to speed up these destructive processes to an alarming degree and to wipe out entire animal communities. In a few cases he has acted more thoughtfully and started to stabilise the various relationships among wild animal populations so that these are once again adapted to their ecosystems. Thus in certain regions he has found native wild ungulates to be a viable alternative to domestic animals. One example is that of the Mongolian saiga, which was practically exterminated by about 1918. However, as a result of an enlightened policy of land management, it today represents one of the principal resources of the local populations.

The populations of many wild species have been drastically reduced by man's encroachment on their natural habitats. There

▲ Statue of a monkey at Singapore

will always be demand for such land for commercial development, but man is gradually beginning to think of the native animals as well as himself. Today there is an increasing awareness of the need for conservation and some efforts are being made to exploit natural resources in a more rational and sensible manner so that both mammals and man will benefit in the long term.

Classification

Taxonomy is the branch of natural history which is concerned with the relationships that exist between different species and the structure of the species themselves. The classification of a particular group should represent the sum total of its evolutionary history, presenting it in the form of a genealogical tree, continuously developing and perhaps throwing out new branches. The individual species are by definition self-contained populations inter-breeding among themselves and isolated, reproductively, from populations of other species. However, the internal structure of such a population may vary, sometimes appearing very complex. Thus it may be divided into subspecies (in which case it is known as polytypic) or not (when it is called monotypic). Therefore, the function of taxonomy is to study evolution by looking at species rather than single cells, tissues and organs. The most important aspects of the evolutionary process itself are the

origins of new species, the ways in which genetic material within those species brings about modifications and the manner in which populations isolate themselves and so preserve their identity.

In animal classification taxonomic schemes group species in genera, genera in families, families in orders and orders in classes. Sometimes intermediate categories introduced for the sake of clarity. In the case of mammals the taxonomic scheme accepted by most scholars today is the one devised by G. G. Simpson in 1945. This is based on a wide knowledge of living and fossil forms and a critical analysis of significant characteristics.

Mammals are divided into three subclasses – **Prototheria**, comprising only the monotremes; **Allotheria**, including a number of extinct groups; and **Theria**, embracing all living orders as well as a few that are extinct. The following list is the recommended classification of the mammals.

▲ Australian echidna (*Tachyglossus aculeatus*)

CLASS MAMMALIA

Subclass Prototheria
Order Monotremata
A very isolated group probably derived from a side branch of synapsid reptiles, in contrast to other mammals, but not known until quite recently in fossil form. It comprises two families: Ornithorhynchidae, with a single species, the Duckbilled platypus; and Tachyglossidae, with five species of echidna of the genera *Tachyglossus* and *Zaglossus*. Unlike other mammals they are oviparous (egg-laying) but feed their young on secretions from mammary glands.

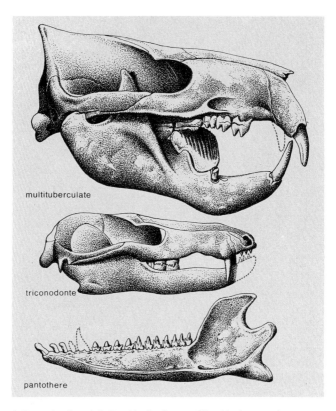

▲ Reconstruction of skull and teeth of some of the oldest mammals

Subclass Allotheria
Order Multituberculata
A group that appeared at the end of the Jurassic, probably derived from a separate side branch, and vanished at the beginning of the Eocene without leaving any descendants. The skull bears a superficial resemblance to that of rodents.

Groups of uncertain rank
Order Triconodonta
Small carnivorous mammals of the Mesozoic, probably derived from a distinct side branch and characterised by teeth furnished with three longitudinally aligned cusps.

235

Subclass Theria

Order Pantotheria or Trituberculata

Probably the ancestors of all later mammals, the mammals of this order are known in fossil form from the Late Jurassic and have characteristic molars with three triangularly arranged cuspids which is the basic pattern of all the Theria.

Order Marsupialia

A group probably descended from the Pantotheria, evolving in parallel to the Eutheria. Appearing in the Late Cretaceous, the marsupials became widespread throughout the world, leaving many fossil remains. Today they comprise two families from South America, Didelphidae and Caenolestidae (the opossum, of the former family, being found in the Nearctic region), and six from Australia which have evolved spectacularly in adaptations to different modes of life. These exhibit a number of convergent and parallel features with placental mammals. The Australian families are the Dasyuridae (Dasyures), Noctoryctidae (marsupial moles), Peramelidae (bandicoots), Vombatidae (wombats), Macropodidae (kangaroos) and Phalangeridae (phalangers, Australian possums and cuscus). Representatives of the Phalangeridae are also found in Celebes and the Solomon Islands.

Order Insectivora

Considered to be the most primitive order of eutherians or living placental mammals, although some forms display strikingly specialised features. Fossils go back to the Late Cretaceous and they are descended from the pantotheres, developing in parallel to the marsupials. From these forms are derived all the other orders of eutherians, some of which are closely related and grouped

▼ White toothed shrew (*Crocidura russula*)

◄ Virginian opossum
(*Didelphis marsupialis virginiana*)

▲ Flying lemur (*Cynocephalus*)

together by Simpson as follows: Insectivora, Dermoptera,
Chiroptera, Primates, Tillodontia, Taeniodonta, Edentata and
Pholidota.

There are 370 known species of insectivores, distributed
throughout the world apart from Australia and New Zealand.
The families are Chrysochloridae (African golden moles),
Solenodontidae (a small family found only in the Antilles and now

almost extinct), Tenrecidae (mainly from Madagascar), Soricidae (Shrews), Talpidae (moles) and Macroscelididae (jumping or elephant shrews).

Order Dermoptera

A group native to the Oriental region, containing two species of the single genus *Cynocephalus*. Arboreal, herbivorous and capable of gliding flight, these flying lemurs have molars similar to those of certain fossils of the Palaeocene and Eocene.

Order Chiroptera

Large order with very many species, known in fossil form since the Eocene, the species being similar in almost every way to those of today. The bats are the only mammals capable of true flight and are diffused all over the world apart from the polar regions. These nocturnal animals are divided into two large groups: Megachiroptera (fruit-eating bats) and Microchiroptera (insect-eating bats). The latter are specialised in the use of echo-location.

▼ Horseshoe bat (*Rhinolophus ferrum equinum*)

239

▲ Common tree shrew (*Tupaia glis*)

The suborder Megachiroptera consists of the single family
Pteropodidae, while the Microchiroptera comprises some
fourteen families with about 650 of the total number of 780
species. These are the Nycteriidae, Megadermatidae,
Hipposideridae (of which the Rhinolophidae are often taken to be
a separate family), Noctilionidae (fish-eating or bulldog bats),
Phyllostomidae (vampire, false vampire and nectar-feeding bats),
Rhinopomatidae (mouse-tailed bats), Emballonuridae (tomb or
ghost bats), Natalidae, Furipteridae, Thyropteridae,
Myzopodidae (confined to Madagascar), Vespertilionidae (the
main family with 275 species and worldwide distribution),
Mystacinidae (with one species endemic to New Zealand) and
Molossidae.

Order Primates
A much studied group which includes the human species. The
order contains a multitude of graduated forms, from types
resembling insectivores to anthropoid apes. Numerous fossil
remains have thrown light on the evolutionary history of monkeys
and of man. This order, widespread through the tropical regions,

240

▲ Chimpanzee (*Pan troglodytes*)

except for Australia, but present during the Tertiary in the
Holarctic region as well, contains 196 living species divided into
three suborders: Lemuroidea, Tarsioidea and Anthropoidea. The
Lemoroidea comprise the Tupaiidae of the Oriental region,
similar to insectivores, the Lemuridae of Madagascar, Indriidae
and Daubentoniidae (also native to this island), and Lorisidae
(lorises and bushbabies). The Tarsioidea consist of the single

▲ Nine-banded armadillo (*Dasypus novemcinctus*)

family Tarsidae of south-east Asia. The anthropoidae are divided into a Central and South American suborder, Platyrrhini, with the families Hapalidae and Cebidae, and the Old World suborder of Catarrhini, with the families Cercopithecidae, Pongidae and Hominidae, the last represented only by man.

Order Edentata
A group that was once abundant, appearing in the Palaeocene and exclusive to South America. The order comprises the sub-orders Palaeonodontidae, with a few fossil forms, and Xenarthra. The latter consists of the infraorder Pilosa (hairy edentates) with the families Myrmecophagidae (true anteaters) and Bradypodidae (sloths), and the Cingulata (armoured edentates) with the family Dasypodidae (armadillos).

Order Pholodiota
A group represented by the single genus *Manis*, with seven species of pangolins from Africa and Asia, some arboreal, some ground dwellers, with a body covered by horny plates and forefeet equipped with strong claws for digging.

▲ Chinese pangolin (*Manis pentadactyla*)   243

▲ Varying hare (*Lepus timidus*) in winter coat

▲ Common hamster (*Cricetus cricetus*)

Order Lagomorpha or Duplicidentata

This order was once considered to be a suborder of the Rodentia because of the superficial resemblance of its members to actual rodents. Among several divergent features going back to ancient times is the presence of a second pair of incisors in the upper jaw. The order is diffused throughout the world except for Australia, Madagascar and various islands, but has become cosmopolitan as a result of man's introduction of certain species, notably the rabbit and hare. There are 64 species, subdivided into two families: Leporidae, including the hares and rabbits, with long ears and short limbs; and Ochotonidae, the calling hares or pikas, with short ears, represented by twelve species in Siberia and the Himalayas as well as two from the Rocky Mountains.

Order Rodentia

This is the largest of all mammalian orders. There are, in fact, more genera, species and individuals of rodents than in any other orders of mammals. Fossils have been found from the Late Palaeocene. Rodents are perfectly adapted for gnawing, with a pair of large, continuously growing incisors in the upper and lower jaws, and a series of molar-type teeth separated by a wide diastema or gap from the incisors. They are essentially herbivorous, either living in trees or below ground. The majority

▲ Cuvier's whale (*Ziphius cavirostris*)

are small in size, although some species are quite large, the capybara being about the size of a small pig. Today there are about 1,900 species belonging to 35 families, divided into three suborders, Sciuromorpha, Myomorpha and Hystricomorpha. The Sciuromorpha are the Aplodontidae (represented only by the mountain beaver), Sciuridae (with some 300 species of squirrels, flying squirrels and marmots), Anomaluridae, Gliridae, Selveniidae (a single insect-eating species from the Asiatic deserts), Heteromyidae, Geomyidae and Ctenodactylidae. The Myomorpha comprise the Zapodidae, Dipodidae, Cricetidae (with 464 species), Platacanthomyidae, Microtidae, Gerbillidae, Muridae (with 463 species) and Pedetidae (jumping hares). The Hystricomorpha consist of the families Hystricidae, Castoridae, Thryonomyidae, Petromyidae, Bathyergidae, Spalacidae, Rhizomyidae, and those native to South America, namely Erethizontidae, Caviidae, Hydrochoeridae, Cuniculidae, Chinchillidae, Dinomyidae, Capromyidae, Echimyidae, Abrocimidae, Ctenomyidae, Octodontidae and Dasyproctidae.

Order Cetacea
A group of obscure origin comprising forms specialised for aquatic life, with many anatomical modifications and a well developed system of echo-location. Among them is the largest

mammal that has ever lived, the blue whale, up to 33 m in length and weighing up to 130 tonnes. Present-day species are to be found in all the world's oceans and some are endemic to lakes and rivers. The suborder Archaeoceti comprised primitive forms from the Eocene and the beginning of the Miocene which are probably not the direct ancestors of living species. The suborder Odontoceti (toothed whales) includes the large majority of extant cetaceans, of varying sizes. The families are the Platanistidae, (freshwater or river dolphins), Delphinidae, Monodontidae, Physeteridae and Ziphiidae. The suborder Mysticeti (baleen or whalebone whales) is made up of the Eschrichtidae, Balaenopteridae and Balaenidae, these being the true baleen whales.

Order Carnivora

This order consists of predatory animals equipped with large canine teeth and a premolar or molar carnassial tooth in the upper and lower jaws, designed for shearing food. The order has existed since the beginning of the Palaeocene, and its present-day representatives differ greatly in life styles, habits and capabilities. There are approximately 290 species belonging to ten families, found all over the world apart from Australia and Antarctica. The order is divided into three suborders, Creodonta, Fissipeda and Pinnipedia. The creodonts became extinct at the end of the Miocene. The fissipeds, descended from a branch of creodonts during the Palaeocene, are characterised by the fact that their carnassial teeth are the last upper premolar and the first lower molar. The pinnipeds, derived in the Late Eocene from a group of

▲ Raccoon (*Procyon lotor*)

▲ The largest predatory land mammal was
Andrewsarchus, a Late Eocene creodont from Mongolia

fissipeds, are specialised for aquatic life even though they
reproduce and suckle their young on land. The fissipeds comprise
seven families: the Canidae, Ursidae, Procyonidae, Mustelidae,
Viverridae, Hyaenidae and Felidae. The pinnipeds are subdivided
into three families: Otariidae, Odobenidae and Phocidae.

Order Astrapotheria
A South American group which flourished between the
Palaeocene and the Pliocene, probably having the same origins as
the notoungulates. Some of these animals were large, rhinoceros-
like species, with enormous canines and a short proboscis.

Order Tubulidentata
Small indigenous African order, comprising the aardvark or
African anteater and a few fossils from Eurasia dating from the
Pliocene. Their molars have no enamel and are covered with
cement, the pulp cavity being formed of a series of tubules.

247

◀ *Uintatherium*, which became
extinct in the Late Eocene

▼ *Astrapotherium magnum*

▲ African elephant (*Loxodonta aficana*)

Order Dinocerata
Large animals similar to pantodonts which lived from the
Palaeocene until the Eocene in the Holarctic region. The best
known genus is *Uinatherium,* with three pairs of horns and
enormous upper canines resembling tusks.

Order Proboscidea
A group nowadays represented only by two species of elephants
but of which hundreds of fossil species from Africa have been
found, dating from the Eocene, as well as from Europe, Asia and
America.

▲ Rock hyrax (*Procavia*)

Order Hyracoidea
A group of small African and Arabian mammals known as rock hyraxes, represented by the single family Procaviidae, already known in the Lower Oligocene.

Order Sirenia
Specialised aquatic mammals who had the same origin as the proboscidians. They are today represented by the family Dugongidae (dugongs), with a single species, and the Trichechidae (manatees) of the African and American Atlantic coasts, with three species.

▼ Dugong (*Dugong dugong*)

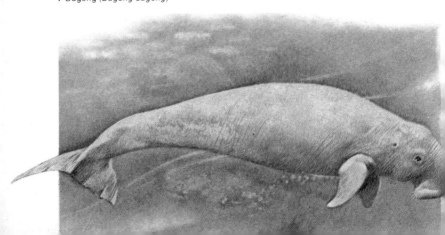

Order Perissodactyla

There are only about fifteen present-day species of perissodactyls – ungulates in which the third toe is hughly developed – but a much larger number flourished during the Miocene and are extinct. Modern species are divided into the suborders Hippomorpha, with the single family Equidae (horses, asses, zebras), and Ceratomorpha, with the families Tapiridae (tapirs) and Rhinocerotidae (rhinoceroses).

Order Artiodactyla

This is an extremely heterogeneous order comprising many living and fossil forms that are divided into various suborders. These are ungulates with an even number of toes, of which the third and fourth are particularly developed. The teeth are specialised for a herbivorous diet. The oldest forms date from the North American Eocene, and the order developed enormously during the Tertiary, with numerous families, some of which are now extinct. The present-day survivors are grouped in three suborders, Suiformes, Tylopoda and Ruminantia. The Suiformes comprise the Suidae (pigs and boars) from Europe, Asia and Africa, the Tayassuidae (peccaries) of South America and the African Hippopotamidae. The Tylopoda are represented by the single family Camelidae, found in Asia and South America. The Ruminantia consist of the families Tragulidae, from Africa and Asia, the Cervidae (not found in the Ethiopian region), Giraffidae (indigenous to the Ethiopian region), Antilocapridae, present only in North America, and Bovidae, not found in South America, with a very large number of species.

▼ Mouflon (*Ovis musimon*)

Bibliography

Bang, P./Dahlstrom, P., *Animal Tracks and Signs*. Collins, London, 1974.

Corbet, G. B./Southern, H. N. (eds.) *Handbook of British Mammals*. Blackwell Scientific Publications, 1977.

Dorst, J., *Before Nature Dies*. Collins, 1970.

Ewer, R. F., *The Carnivores*. Cornell University, Ithaca, 1973.

Frisch, D. von, *Animal Migration*. Collins, London, 1969.

Griffin, D. R. *Echoes of Bats and Men*. Anchor Books, New York, 1959.

Grzimek, B., *Animal Life Enclyclopedia* (13 Vols.). Van Nostrand Reinhold, New York, 1972–1975.

Lawrence, M. J./Brown, R. W., *Mammals of Britain – Their Tracks, Trails and Signs*. Blandford, London, 1973.

Matthews, L. H., *Life of Mammals* (2 Vols.). London, 1960–1971.

Mayr, E., *Animal Species and Evolution*. Harvard University Press, Cambridge, 1963.

Norris, K. S., *Whales, Dolphins and Porpoises*. University of California, Berkeley, 1966.

Romer, A. S., *Vertebrate Paleontology*. University of Chicago, 1966.

Schultz, A. H., *Life of Primates*. Weidendfeld and Nicholson, London, 1974.

Van Den Brink, F. H. *Field Guide to Mammals of Britain and Europe*. Collins, London, 1967.

Walker, E. P., *Mammals of the World* (Second Edition, 2 Vols). Hopkins, Baltimore, 1975.

Young, J. Z., *The Life of Mammals*. Oxford University Press, 1970.

Index